Celebrating Soil

M.R. Balks • D. Zabowski

Celebrating Soil

Discovering Soils and Landscapes

 Springer

M.R. Balks
University of Waikato
Hamilton, New Zealand

D. Zabowski
University of Washington
Seattle, WA, USA

ISBN 978-3-319-81347-9   ISBN 978-3-319-32684-9   (eBook)
DOI 10.1007/978-3-319-32684-9

Cover page: Van Gough looks deeper – tapestry in wool by Megan Balks

Printed on acid-free paper

This Springer imprint is published by Springer Nature
The registered company is Springer International Publishing AG Switzerland

Preface

Our objective is to produce a book that celebrates the diversity, importance, and intrinsic beauty of soils around the world and helps the reader to understand the ways that soils are related to the landscapes in which they form and the life that they support. Jock Churchman in 2013 nicely captured our underlying philosophy when he wrote: "Like the Lorax who speak for the trees in Dr Seuss' wonderful environmental fable, one of the most important roles for *soil scientists* may be to speak for the soil." We want every person on Earth to appreciate how important our soil resource is. Swaisgood and Sheppard in 2010 suggested that "The gloom and doom niche in conservation is well occupied, ... By contrast the hope niche is relatively open...." Thus we want to give a positive message, rather than the fearful warnings that tend to dominate much of today's environmental writing.

We want to take our readers on a global journey through soils and landscapes ranging from the Arctic to the Antarctic, sharing the delight of discovery and sense of awe for the amazing variety and processes that soils utilize to support the ecosystems of different landscapes. Exploring the relationships between landscapes and the underlying soils can increase a reader's understanding of the soils beneath the landscapes in which they live, work, and travel.

A book of this type cannot be comprehensive. The soils and landscapes discussed are ones in which at least one of the authors has first hand experience. By using a wide range of examples, we seek to give some insight into the variety and fascination that is to be found in the study of soils. There are many excellent publications and websites people can go to for more in-depth information. A list of sources and some suggested further reading are included in Appendix 2, as a starting point for the interested person.

The best way to learn and get to know soils is to get out there and get close to the soil. Plant a garden, make a clay sculpture, and enjoy the feel, sight, sound, and smell of the soil in its endless dance with water and life.

Hamilton, New Zealand
Seattle, WA, USA

M.R. Balks
D. Zabowski

Acknowledgments

We would like to thank:

- University of Waikato in New Zealand and University of Washington, in Seattle, USA, who allowed the authors time to work on this project.
- The many wonderful soil scientists who have provided us with inspiration, friendship, and education over the years. Megan would especially like to thank Vince Neall, Jim Pollock, Allan Hewitt, John McCraw, Iain Campbell, Graeme Claridge, Daniel Hillel, and Sergey Goryachkin. Darlene would like to thank all of the soil scientists who have shared their knowledge and experience with her over many meetings, discussions, and field trips.
- David Lowe, Russell Briggs, Tanya O'Neill, and Wendy Fallon for critical review of drafts of the manuscript.
- Marianne Coleman for assistance with the preparation of figures and diagrams.

Special thanks go to our wonderful husbands, Errol Balks and Brandon Cole, who have provided endless support to us as we have pursued our interests in soil science and in preparing this book.

Contents

Chapter 1
Soils in Harmony with the Environment

Productive landscape in the Alsace region of France with cropping on the fertile, near-flat soils in the valley floor, perennial crops of grapes on the lower hill slopes, and forest on the steeper slopes protecting the shallower hill soils from erosion

© Springer International Publishing Switzerland 2016
M.R. Balks, D. Zabowski, *Celebrating Soil*, DOI 10.1007/978-3-319-32684-9_1

Introduction

"The source and final resting place of everything that grows, soil thus inspires reverence not only in the peasant who derives his daily bread from it, but also in the scientist who contemplates its meaning as the place where life and death meet and exchange vital energies"

Daniel Hillel 1991

Soil, like air and water, is critical to terrestrial life on Earth. Soil underpins human food supply and provides materials on which we build our lives. Soil is out of sight and often out of mind, thus easy to overlook. Yet soil has tremendous variety and intrinsic beauty for those who take time to look (Fig. 1.1).

Soil is the top meter or so of unconsolidated material at the Earth's surface formed where air, water, and life interact with geological materials. Soils contain a memory of events that have shaped the landscape and the environment. With a little knowledge you can look at a soil and understand some of the stories that it has to tell.

Our planet contains a wonderful diversity of landscapes from the cold deserts of Antarctica to the hot deserts of North Africa, the high mountains of the Andes in South America to the rainforests of the Pacific Islands. Each landscape has a different set of soils developed upon it, and within each landscape there is variety and complexity in the soils that form.

Fig. 1.1 Two contrasting soils from opposite ends of the Earth with different stories to tell. (Markers on tapes are 10 cm intervals.) *Left*: an unexpectedly red soil from near the Arctic Circle in Russia reflects the red rocks from which it has formed. The white layer near the surface is the result of acids washed in from the overlying forest vegetation causing the iron-rich red minerals to leach down through the upper part of the soil, leaving only a white sand behind. *Right*: a particularly odoriferous soil formed in penguin guano, and the stones from which penguins make their nests, in Antarctica

Soils are buried treasure and a journey exploring soil provides a wealth of knowledge and experience. There is pleasure to be had in such a treasure hunt: in the food and flowers that grow in soils, in the joy of seeing the colors and patterns that occur, in the sensuous pleasure of working clay in your hand or feeling the warm earth beneath your feet, and in the rich earthy scent of soil after warm rain.

The Environment and Soil

Soils develop in response to, and in harmony with, the environment. Soils are never static as the constituents from which a soil is formed slowly weather and are altered, new minerals are formed, and materials are gradually added or removed. Processes in soils are often imperceptibly slow, by human standards. However soils have time on their side. Occasionally the life of a soil is punctuated by sudden events, such as floods or landslides (or in this day and age attack by bulldozers) that may cause great changes. Plants and animals add, remove, and move materials within soils. Rainwater moving through soil, along with summer heat and winter frost, impacts on the soil and the organisms that dwell in it. The differences between different soils can be explained in terms of a key set of five environmental factors (Fig. 1.2). The five "soil-forming factors" are:

- The geological materials and organic matter that form the *parent materials* from which the soils are derived
- The influence of *topography* or position in the landscape
- The effects of *climate* (primarily rainfall and temperature)
- The impacts of living *organisms*—the myriad plants, animals (including humans), and microbes that live in, or depend on, the soil
- The *time* that the soil has had to develop

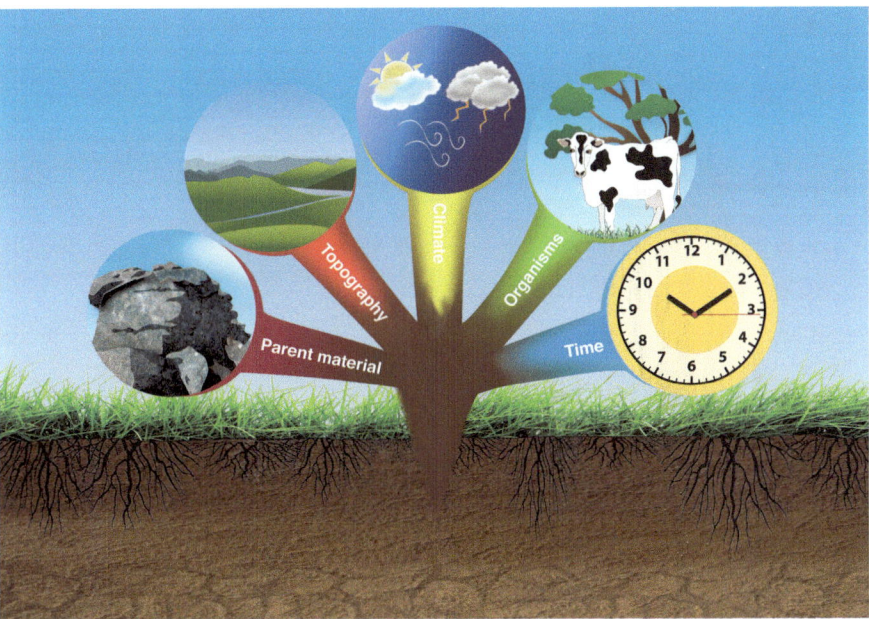

Fig. 1.2 Five environmental elements—the soil-forming factors—interact to create an infinite variety of soils. Soils evolve, over time, in response to the environment

All of the environmental factors interact with one another creating the infinite variety that exists in our soils and the diversity of soils that we find in our landscapes.

In this chapter we will explore each of the soil-forming factors and the variety of soils that evolve in response to variations in the environments to which soils are exposed. Some of the key concepts of the soil and its role in supporting the growing human population are also discussed.

The Underlying Sources of Soil: Parent Materials

Soils are derived from the geological and organic materials in which they form (the parent materials). The diversity of geological materials impacts the resulting soils. Hard, silica-rich rocks, such as quartz sandstones, often weather slowly. Softer, fine-textured rocks, such as weakly cemented mudstones, may weather faster. Different rocks contain different minerals and thus make different chemical elements available to the soil.

Many soils are not formed solely in solid rock, but rather in materials that have been moved around the landscape such as sand dunes; silt, sand, or gravel deposited on a river floodplain; or volcanic material carried through the air. Each material leaves its imprint on the resulting soils. The range of plant nutrients available in a soil mainly depends on the elements that were available from the parent materials. Some rocks, such as limestone, have a high pH, which is inherited by the soil, while others such as a quartz sandstone will usually weather to an acidic (low pH) soil.

Some soils contain minimal rock material as they form from organic, usually plant, material. In swampy, water-saturated environments, dead plant material gradually accumulates, rather than decomposing. Deep deposits of organic matter form peat and the resultant soils tend to be acidic and low in nutrients.

The Russian subarctic boreal forest, where spruce and fir trees dominate the landscape, provides us with an example of the effects of soil parent material on soil properties (Fig. 1.3). The climate is harsh

Fig. 1.3 Soils from the forest zone of northern Russia. Differences between the soils are due to the different parent rocks on which they are formed. *Left*: the landscape near the Pinega River. *Center*: soil formed in sandstone with a typical white layer due to acidified water seeping down from the forest litter removing the reddish-brown iron minerals (podzol). *Right*: soil formed in limestone where the soluble rock disintegrates leaving only an organic matter-rich topsoil over yet to be weathered rock (rendzina). Soil depths are shown in cm

with short, warm summers and long, dark, snow-covered winters. Here we find a surprising diversity of soils, due to the varied underlying geological materials. A soil called a "podzol" with a distinct white, acid-leached layer near the surface is the dominant soil that we expect to form on most parent materials in the Russian forest environment. However, we can contrast the podzol with the neighboring soil formed on limestone rocks where the distinct black topsoil overlies white rock. Limestone is soluble in water and distinctive soils, with a relatively alkaline (high) pH, and minimal B horizon, form on limestone parent materials.

Topography: The Relationship Between Landscape and Soils

The location of a soil within the landscape influences the type of soil that forms (Figs. 1.4 and 1.5). The topography of a site describes the shape of the land including the steepness of the slope, the aspect or direction the slope faces, and the elevation of the site. Understanding the relationship between the topography, of a landscape, and the other factors that influence soils enables soil

Fig. 1.4 A mountainous landscape in Switzerland. Shallow rocky soils occur near the mountaintops. Deeper soils form on eroded materials deposited as fans at the foot of the slopes and on river deposits in the valley floors

Fig. 1.5 A landscape and associated soils in the Waitetuna Valley of New Zealand. Soil *A* is formed on the floodplain deposits of the small stream in the foreground. The soil is fertile due to additions of new silt during regular floods. It is saturated with water in the lower part of the profile due to the proximity of the water table. Soil *B* is on the gently rolling slopes of the lower part of the hills. Here tephra (volcanic ash) from distant eruptions has accumulated over thousands of years giving the soil its distinctive color and excellent soil physical properties. Soil *C* is formed on the steep hills in the middle distance. Here any tephra that falls is eroded off, and so the soil is a product of weathering of the underlying rocks. Soil *C* has a high clay content, is strongly leached, and is not very fertile. Soil *D* is formed on the high slopes of the distant mountains where the altitude leads to cooler temperatures and stronger erosion from high winds and rainfall. The soil is shallow and stony with only stunted shrubs, mosses, and lichens surviving

scientists to develop soil-landscape models that allow them to accurately predict the properties of the soil in different parts of the landscape.

Slope, elevation, and aspect all impact on the climate experienced at a particular site in the landscape. In the Northern Hemisphere, soils on a south-facing slope will usually be drier and warmer than those on a north-facing slope and vice versa in the Southern Hemisphere. The climate is always cooler at higher elevations—for every 100 m gain in altitude, the temperature drops about 1 °C. Higher elevations often experience stronger winds and more intense rainfall. Slope position influences the properties of a soil relative to the other sites around it. A landscape may also have micro-topography such as small mounds or hollows that may only be a meter in diameter, but on which the soils that form are distinctly different from their neighbors.

Topography can influence a variety of soil processes, but often the most important effects are related to water flow, temperature, and soil movement. Soils at the base of a slope or in a valley are more likely to be wetter than those on a ridge or upper slopes due to water moving downhill and persisting at lower sites. If a soil is near the edge of a ridge or located on a steep slope, it is also likely to be shallower (particularly the surface horizons) than a soil near the toe of the slope or in a valley. Soil is more likely to be eroded from ridges and hilltops, and the eroded soil often accumulates near the base of the slope making valley floor and toe-slope soils deeper.

Topography and parent material often change together. For example, a valley may be filled with alluvial deposits from a river, while a hill has soil developing from bedrock. Likewise, topography can influence localized differences in climate, vegetation, and other organisms. Downslope soils are often wetter due to runoff and are more shaded by adjacent hills, so are cooler, and have different vegetation from the warmer, drier, upland soils.

Vegetation often varies within a landscape as it is influenced by the variations in shading, moisture availability, and other factors that are influenced by topography. Vegetation in turn also impacts on the soils that form. At higher altitudes where vegetation struggles to survive in the harsher environment, the absence or limited presence of vegetation means there is less protection of the soil from erosion. Human activity can also strongly impact on vegetation and soils within the landscape. For instance, flatter, low-lying areas are much more likely to be subjected to plowing and establishment of crops, whereas higher steeper slopes are more likely to be left for forests to develop.

Climate: Hot or Cold, Wet or Dry, Soils Respond

Climate influences both temperature and water availability and thus the soil that forms. Temperature controls the rate of chemical reactions, such as the reactions that convert hard rock minerals into soil clays. As a rule of thumb, within the temperature range commonly found in soils, for every 10 °C rise in temperature, the rate of biochemical reactions will double. Thus soil chemical weathering operates faster in warmer climates as do most biological processes. Temperature extremes, both hot and cold, limit the ability of plants, animals, and most microbes to survive in the soil.

Water is critical to many soil processes. Often the chemical reactions that break down rock minerals involve reactions with water. Water is vital to the growth and survival of plants and soil organisms. Water also influences soil physical processes. Much soil erosion is facilitated by water. When water freezes in the soil, it expands forcing soil aggregates apart and helping to break up rocks. Nutrients in the soil can dissolve in water and can then be taken up by plant roots or washed down, through the soil, into the groundwater and then ultimately into rivers and the sea, thus making the sea salty.

The broad climate zones of the planet (Fig. 1.6) are generally associated with a suite of soil properties. In the warm, wet tropics, soils tend to weather strongly to deep, relatively infertile, red-colored, clay-rich profiles. In the low-latitude deserts, such as the Sahara, water limits plant growth and soil

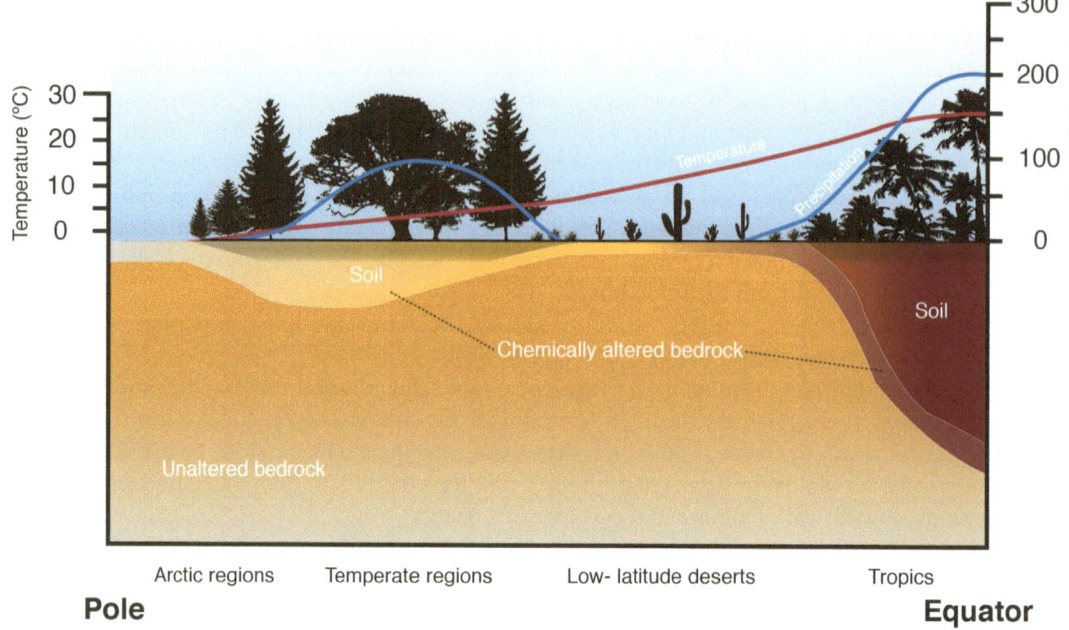

Fig. 1.6 The pattern of climate and depth of weathering and soil development from the equator to the poles

Fig. 1.7 A deep, red-weathered soil from the warm wet tropical environment of Thailand supports good crop growth when plant nutrients are added

development, resulting in sandy desert environments. In temperate regions moderate soil development supports productive grasslands and forestlands. In the cold desert of the Antarctic, chemical weathering is limited, and soils mainly comprise gravelly sandy materials broken down from solid rocks by the actions of glaciers, wind abrasion, and water freezing and thawing.

The tropical heat of Thailand provides us with an example of the effects of a warm wet environment on soil development (Fig. 1.7). In tropical regions the weathering of soil materials to form clay extends tens of meters beneath the soil surface. However, tropical soils are often poor in plant

Fig. 1.8 A relatively shallow stony soil from the cold desert of the Wright Valley in Antarctica where life is generally limited to microbes, and soils contain minimal clay or organic matter

nutrients as soluble products of weathering (such as potassium and magnesium which are important plant nutrients) have been leached from the soil by the frequent heavy rainfalls. In the warm wet environment, organic matter from dead plants or animals is quickly decomposed, and the nutrients are taken up by the vegetation or leached away by the rain. Thus in tropical forests, the vegetation, rather than the soil, contains most of the plant nutrients. If tropical forests are removed, the nutrients are often lost with them, leaving a further impoverished soil environment.

In contrast, in the Antarctic Dry Valleys, we find some of the coldest, driest soils on Earth (Fig. 1.8). Here chemical weathering proceeds extremely slowly with just a faint tinge of rust color showing that some iron oxides have formed. The soil is dominated by rocks, gravel, and sand that have formed due to physical abrasion of the rock materials when they were moved and deposited by glacier ice. Soils in the Dry Valleys contain a diversity of cold-tolerant microbes, but no higher plants survive, so organic matter in the soils is negligible.

Living Creatures: Influence on the Soils on Which They Depend

A myriad of organisms live in the soil and all terrestrial life ultimately derives its sustenance from the soil. In the natural world, when an organism dies, its remains are broken down by soil microbes and thus its energy and nutrients are harnessed to support a new generation of life.

A teaspoon of topsoil may contain as much as 100 billion bacteria and 15 km of fungal hyphae along with a host of other microbes, ranging from tiny worms, called nematodes, to microscopic insects such as springtails (Fig. 1.9). Larger microbes such as protozoa dine on the bacteria, and tiny animals such as rotifers may in turn eat the protozoa. Thus the food chain begins. Many soil microbes dine on dead plant and animal matter and, in the process, release nutrients. Plants can then take up the nutrients. Thus the activity of soil microbes provides a vital recycling service that underpins the whole food chain.

Many larger insects and animals make their home in the soil. For example, earthworms consume organic matter and eat soil mineral material with it; after they digest the lot, their waste is often

Fig. 1.9 Soil organisms.
Top: a scanning electron
microscope photo allows
us to see soil microbes.
The shell creature is a
"testate amoebae," a soil
protozoa that eats bacteria.
The tiny rectangular shapes
below the protozoan are
bacteria that, along with
the slime they produce, are
coating the soil surface.
(The dotted scale is
12/1000 s of a mm long.)
Middle: earthworms are a
common soil inhabitant.
Bottom: a termite mound
in Australia demonstrates
the ability of some insects
to move soil materials

deposited as worm casts at the soil surface. Worms also create burrows that allow air and water to move readily and rapidly through the soil. Termites, wasps, ants, moles, and rabbits are all examples of creatures that move soil material around in order to create a suitable home.

As plants grow and die, they provide a vital source of organic matter for soils. Some plants provide surface leaf litter, that is, a source of nutrients and also a protective layer that accommodates many small insects and microbes, as well as helping capture moisture and reducing soil erosion. Plant roots explore deep into the soil, extracting moisture and nutrients and helping hold the soil together—thus also preventing soil erosion. Plants often produce acids that are washed into the soil and impact on many of the chemical processes that occur in soils. In saturated environments, when plants die, microbes may not readily break down the organic material, and thus it gradually accumulates to form organic soils known as "peat."

We should remember too that of all the organisms that impact on soils, humans are capable of some of the greatest impacts.

Time: It Takes Time for a Soil to Develop

As the saying goes "all good things take time" and forming a soil can be a very slow process. Soils progress from new material being exposed at the soil surface through to development of a strong and productive soil over thousands of years. Eventually the weathering processes continue until a soil becomes old, weary, leached of nutrients, and weathered to clay. In many parts of the planet, soils are reinvigorated from time to time by processes such as erosion or deposition of new materials. In the relatively rare cases where land surfaces have remained stable for millions of years, the ancient soils that form are beautiful and they contain a record of the past.

The rate that a soil develops depends on other environmental factors such as climate and parent materials. In the extreme cold and dry environment of Antarctica, soil processes operate exceptionally slowly. However some Antarctic land surfaces are thought to be millions of years old, and so even exceedingly slow soil development can proceed to form distinctive soils. In warm wet climates where vegetation thrives, a soil will form relatively quickly, especially if the substrate is an unconsolidated material such as silt or sand.

The youngest soils, often referred to as "recent soils," are formed on newly deposited materials and show only the very first stages of soil development. New materials may include recently deposited lavas (Fig. 1.10), silt deposited by rivers on floodplains, and surfaces newly exposed to soil processes as a result of erosion and removal of the preexisting soil.

Fig. 1.10 On lava erupted in Hawaii 40 years ago, the soil is only just beginning to form. In deeper cracks in the rock, windblown leaf litter has accumulated, providing some organic material. Hardy plants find a niche in the cracks where moisture can be trapped, and there is some protection from the scorching heat of sun on the black rock surface. As more plants colonize the surface, they help enhance the rock-weathering process, and so, over thousands of years, a new soil will form

Fig. 1.11 Shallow sandy soils formed on ancient rocks in Australia in an arid environment where weathered soil material has been periodically removed by erosion

Ancient rock surfaces in Australia have been exposed to many cycles of weathering and erosion. The rocks exposed in the Karijini National Park in Western Australia were formed in the sea about 2.5 billion years ago, and the present-day land surface was formed over the last few tens of millions of years (Fig. 1.11). Ongoing erosion and a harsh, hot, dry climate means that even though the rocks and landforms are ancient, the soils, although much older than those in many parts of the world, are often relatively thin and stony.

Looking Deeper: Understanding the Soil Profile

Factors that affect soil such as climate and organisms have more impact near the surface of the earth. Thus, soil materials weather and change more quickly at the top. To really understand a soil, you should look at a cross section from the ground surface right down to where the original geological material has not been greatly altered. The vertical cross section of soil is called a "soil profile." Looking at a soil profile, you will see that changes in color and other soil properties, such as clay content, occur down the profile. Such changes help us distinguish different zones within the soil which are called "soil horizons" (Fig. 1.12). There are five main soil horizons. A, B, and C horizons can be found in almost any type of soil. The O and E horizons are more common in forested or wetland soils.

A Horizons

A horizons are what most people call top-soil. A horizons have a dark colour due to the accumulation of decomposed organic matter. The A horizon is usually the most fertile part of the soil and it is also typically the most weathered and altered soil if you compare it to the rocky material the soil is developing in.

B horizons

B horizons are often called subsoil. B horizons are not as enriched in organic matter as A horizons, but the parent material has been altered so much that it no longer looks like the original rock material. Usually a B horizon has paler colours; often yellow, orange, or reddish hues. Another characteristic that distinguishes a B horizon is that particles are often clumped into aggregates. B horizons may also have accumulations of materials such as salts, oxides, or clays. B horizons are not usually as fertile as A horizons.

C horizons

C horizons are the underlying, relatively unaltered, rock or parent materials. Soils that are young (or have barely begun to form) may have only a C horizon. If you look closely at a C horizon, you will see bare, unweathered, rock material.

O horizons

Organic matter that accumulates at the soil surface, from falling litter from trees or deposition of decaying vegetation creates an O horizon. The O horizon is composed entirely of organic matter without any mineral soil. O horizons can range from fresh plant material, such as leaves and branches, to well-decomposed and unrecognizable humus.

E horizons

A soil horizon, just below the topsoil, that is white or pale grey may be an E horizon. E horizons have lost organic matter, iron oxides or other materials that give the soil colour, through leaching. E horizons are often found beneath O horizons, but may be underneath an A horizon as well.

Fig. 1.12 Soil profiles with horizons marked. *Left*: a simple soil profile containing A, B, and C horizons. *Right*: a more complex soil profile with O, E, and two B horizons above a C horizon

Celebrating Soils in Harmony with the Environment

Every soil, if studied carefully, like every human face, has etched into it the effects of the environment in which it has developed. Most soils are old, much older than our human timescales allow us to readily appreciate. Some soils have endured through, and often been profoundly altered by, past climate changes. The last glaciation that occurred between about 100,000 and 15,000 years ago had major impacts on the soils and landscapes of all the higher latitude lands such as Europe and North America. Our soils have, over time, been subjected to effects of falling and rising sea levels, the arrival of humans, changes in the vegetation cover, erosion and deposition of materials, and in many cases the development of human agriculture. The soil is quietly resilient, and for all the changes that have occurred, it continues to support the myriad life-forms that make up our ecosystems. It is interesting to take but one example of a soil and look carefully to see what it can tell us (Fig. 1.13).

Fig. 1.13 This soil from Tasmania, in Australia, tells of gradual earth movement on a hill slope which humans have periodically impacted over thousands of years. The soil profile was exposed as the result of a landslide that removed the soil that was adjacent to it. The presence of "improved" pasture grasses tells of human actions—repeated fire may have removed the previous forest vegetation. Fertilizer, along with grass seed, has been added. Roots are evident in the top, overhanging edge of the soil—some left hanging as the soil has eroded away from beneath them. Small, hard, sharp-edged stones have been found in the surface of these soils, possibly left behind by aboriginal peoples long before the arrival of European settlers in the region. The cracks tell of drying clay hint at occasional wet periods when water moves down the cracks bringing the gray sandy surface soil with it until the wetted soil swells again and the cracks are filled. The soil is on a slope and has formed from material that has gradually moved downslope. Within the subsoil, below this picture, there is evidence of a buried soil that once formed the land surface at this site

Chapter 2
Soils Born of Fire

Erupting volcanoes create new land surfaces on which new soils develop, Kiluaea, Hawaii

© Springer International Publishing Switzerland 2016
M.R. Balks, D. Zabowski, *Celebrating Soil*, DOI 10.1007/978-3-319-32684-9_2

Introduction

"Nature is often hidden, sometimes overcome, seldom extinguished"

Francis Bacon, 1625

Volcanoes are one of Earth's most fearsome exhibitions of power. Erupted lavas and tephra (volcanic ash) are a source of materials that weather to form new, and often exceptionally productive, soils.

Volcanoes create and shape landscapes. Whether building huge shield volcanoes, such as the Hawaiian Islands, or coating landscapes with far-flung pumice or tephra from explosive eruptions, such as Crater Lake in Oregon, USA, the effects, and products, of volcanic eruptions are a source of fascination.

People who live in the paths of volcanic eruptions understand the terror of blasts of hot fiery materials or mudflows that extend far beyond the confines of the volcano itself. Mud falling on homes and fields, the inconvenience of airplane flights canceled, and the surreal beauty of bright red sunsets all occur due to volcanic ash, aerosols, and gases, impacting audiences at ever greater distances from the source.

Soils in regions impacted by fallout from volcanic activity contain a memory of the eruptions that have occurred (Fig. 2.1). Burial by volcanic deposits may capture moments in time that are a record of life's tragedy, for instance, the chilling images of people in Pompey, Italy, buried by the eruption from Vesuvius in 79 AD.

Fig. 2.1 This soil, formed in volcanic deposits in the Gunma prefecture of Japan, contains the memory of a long history of volcanic activity. The topsoil includes scoria from the 1783 eruption of Mt. Asama, some 80 km away. The white pumice was erupted from Mt. Haruna about 1400 years ago. The dark-colored buried topsoil contains evidence of cultivation having occurred at this site about 1500 years ago prior to the pumice eruption. The bright brown gravelly layer toward the base of the profile was erupted from Mt. Asama about 15,000 years ago. Below the bottom of this picture is a distinctive tephra erupted from the Aira caldera, about 1000 km away, about 30,000 years ago

Volcanic Landscapes

In order to understand the soils formed from volcanic materials, we need to know something of the volcanic landscapes in which they occur. There are three main kinds of volcano: basaltic, andesitic, and rhyolitic. Each has a distinctive mineral composition and eruption style and each forms a different array of landscapes and soils. Volcanoes are generally associated with tectonic plate boundaries (Fig. 2.2). However, there are also volcanoes, such as the Hawaiian Islands, that are on "hot spots," within tectonic plates, where molten rock from the mantle flows up to the earth's surface.

Basalt volcanoes erupt a dark-colored magma that is rich in iron and magnesium with less than 50 % silica. Basaltic magma rises rapidly to the surface from the Earth's mantle. Basalts are the hottest magmas and generally erupt as very fluid lavas or as fire-fountain eruptions that fling magma into the air, where it cools, solidifies, and lands as basaltic scoria. Basalt is erupted at spreading tectonic plate margins, most of which are beneath the sea and at intraplate hot spots. The very hot, fluid, basaltic lava may flow out over long distances to form huge, low-sloping "shield"-shaped volcanoes (Fig. 2.3). Where scoria is erupted in "fire fountains," it forms small cones, often on top of a larger shield volcano.

Andesite volcanoes form on tectonic plate boundaries where one plate is being subducted beneath another. As the subducted plate dips deep into the Earth's crust, it gets hot and some of the material melts and returns to the surface erupting as andesite. Andesite volcanoes have an "intermediate" composition (with 50–70 % silica) being a combination of material from the Earth's silica-rich surface crust and the iron-rich mantle. Andesite volcanoes tend to alternate between explosive eruptions that blast tephra into the air, and lava flows of viscous magma that do not spread very far. Thus andesite

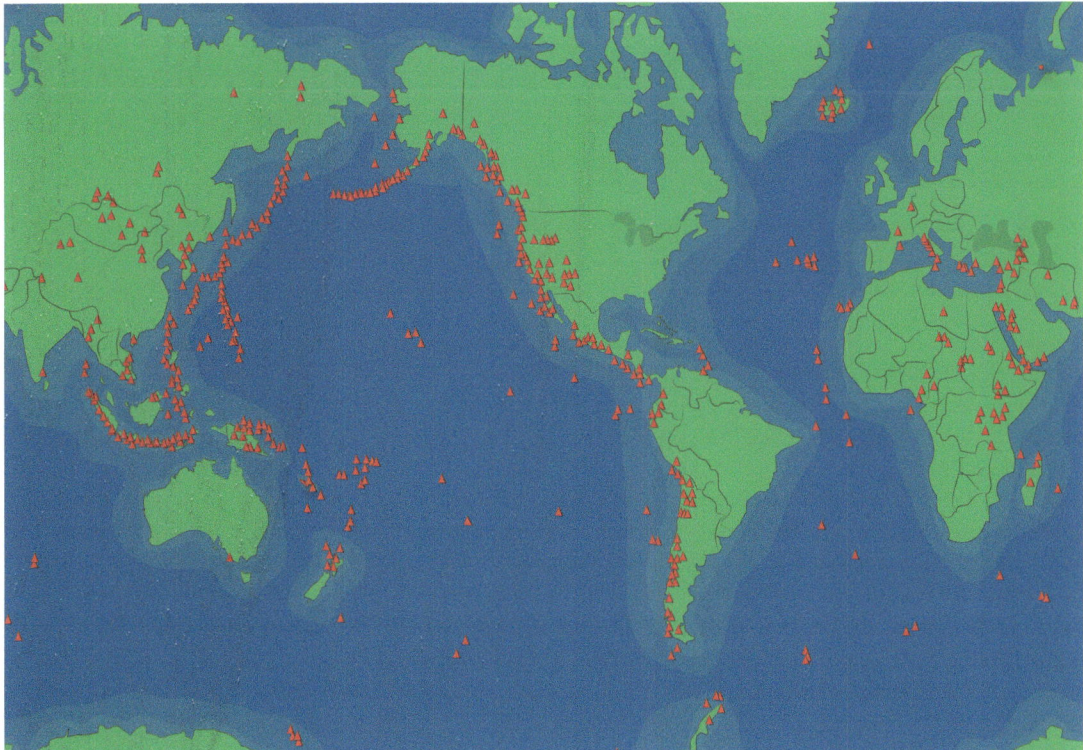

Fig. 2.2 Global distribution of volcanic activity

Fig. 2.3 The low-sloping profile of a basalt shield volcano, Mauna Loa, Hawaii

Fig. 2.4 Mt. Fuji in Japan has the characteristic cone shape that we associate with andesite volcanoes. Fuji has erupted both basaltic and andesitic materials (Photo: David Lowe)

volcanoes build up layers of magma and volcanic ash to create beautiful cones such as Mt. Fuji in Japan (Fig. 2.4).

Many of the steep-sided andesite volcanoes have a large surrounding "ring plain," such as the one around Mt. Taranaki in New Zealand (Fig. 2.5). The ring plain is built up from the layers of tephra that fall from the air and also from rapidly moving, water-saturated mudflows (called lahars) that periodically flow down from the upper slopes of the volcano. As the slope flattens, near the base of the volcano, the lahar flows lose momentum and the material comes to a halt. A hummocky landscape forms which, over time, is colonized and stabilized by vegetation.

Rhyolite volcanoes erupt magmas that are formed from melting of silicic crustal material on subducting tectonic plate margins. Rhyolite magma is high in silica (>70 %) and extremely viscous. Huge pressures build up which, when finally released, result in the largest, most explosive eruptions on Earth.

When rhyolite volcanoes erupt, they eject vast volumes of material high into the atmosphere in a huge eruption column. In some cases the column collapses and hot pumice-rich, gas-laden material is blasted across the landscape, as a fast-moving, ground-hugging, pyroclastic flow that buries or

Fig. 2.5 The ring plain of Mt. Taranaki in New Zealand. The circular change in color around the mountain is the boundary of the Egmont National Park where native forest gives way to dairy pastures. A radial drainage pattern is evident as rivers all flow outward from the top of the mountain, across the ring plain

Fig. 2.6 Crater Lake in Oregon is a caldera formed by collapse of the land following a rhyolitic eruption of about 50 km^3 of material about 7700 years ago

removes everything in its path. Extensive pumice deposits are emplaced, and where the material is hot enough, it fuses together to form a rock called welded ignimbrite. The finer tephra material can be carried thousands of kilometers before settling to the Earth's surface.

Krakatoa, an Indonesian island that erupted in 1883, generated one of the largest historically recorded rhyolite explosions. The blast was heard over 4000 km away and tephra, aerosols, and volcanic gases reached up to the stratosphere, causing global climate cooling which lasted about 5 years.

Once a rhyolite eruption ceases, the ground from which the material was erupted collapses, leaving a huge hole, called a caldera. Often the resulting caldera fills with water to create a lake, for example, Crater Lake in Oregon (Fig. 2.6).

Hawaii: Steam Heat and New Rock

Nowhere else on earth, except perhaps Iceland, is the fiery birth of volcanic soils more evident than in the Hawaiian Islands. People travel from all over the world to see lava flowing from the erupting craters of Kilauea (Fig. 2.7). The Hawaiian Islands sit over a hot spot. Over time the ocean floor has gradually moved across the hot spot giving rise to a series of volcanoes of increasing age with increasing distance from the present-day activity (Fig. 2.8). Thus the oldest eruptive materials (erupted about five million years ago) are found in Kauai, with the youngest material, likely to be erupting as you read this, on the Island of Hawaii.

In the warm, wet climate on the east side of Hawaii (the "Big Island"), it is not long before the newly erupted rock material begins to weather, plants gain a toehold, and soil development gets underway. In the drier western regions, and in the colder high-altitude zones, soil development is slower. The large range in the age of the rocks, the climate, and the vegetation in the Hawaiian Islands leads to a wide variety in the soils that form.

Thus the Hawaiian Islands contain not only a record of the formation of new rock material but also a sequence of soils, ranging from the first steps toward soil formation in the newly erupted material to soils that have developed over millions of years. The new lava flows cool to form hard rock, and their dark color means they heat up in the sun, making the rock surface particularly inhospitable. Where the lava flows have cut through preexisting forests, there is a ready source of nearby seeds and so the plants gradually colonize the new lava (Fig. 2.9).

At first just a few pioneer plants become established, but gradually vegetation gains a hold, the rocks start to weather and more plants are able to survive. After a few thousand years, where the climate is warm and wet, rocks weather, clays are formed, and tropical forest becomes established (Fig. 2.10). Where the climate is dry, weathering is slower. Salts accumulate in the soils and the vegetation is prone to fire damage ensuring survival of only the toughest plants (Fig. 2.11).

Fig. 2.7 New lava flowing from an eruption vent on Kilauea volcano in Hawaii (Photo: Annette Rodgers)

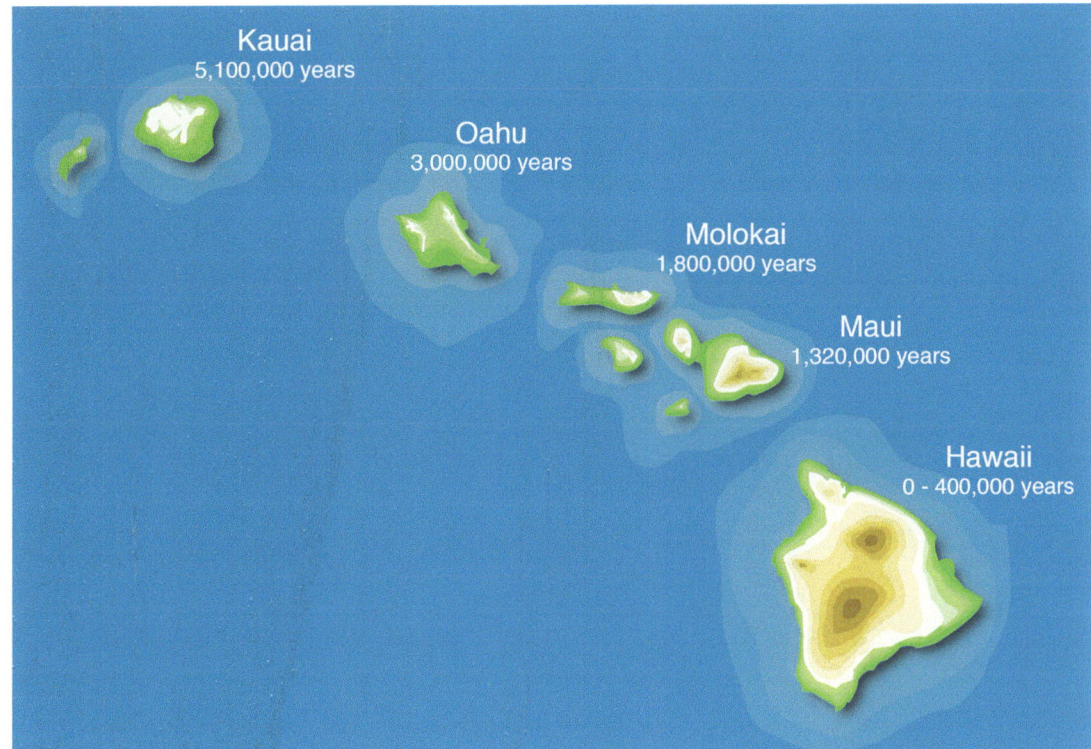

Fig. 2.8 Map of Hawaiian Islands showing ages of volcanic activity

Fig. 2.9 Basalt formed from lava that was erupted from Kilauea in 1973. In these photos taken 40 years after the eruption, the very start of soil development and plant establishment is occurring. The larger, distant trees that survived the eruption provide a source of seed and organic matter to help start new soil formation

Fig. 2.10 Tropical rain forest and the related clay-rich soil formed on ancient lava deposits in the high rainfall on the east side of Hawaii (the "Big Island")

Fig. 2.11 The dry, fire-prone environment and relatively weakly weathered, salt-rich, rocky soil on lava on the west side of the Island of Hawaii

Mount Saint Helens

The Cascade Mountains, in the northwestern USA, are part of the "Ring of Fire" that surrounds the Pacific Ocean. The Cascades have numerous volcanoes, and Mt. St. Helens was one of the most scenic and symmetrical volcanoes with its lower slopes blanketed with forests (Fig. 2.12). Tribal legends called the mountain "Loo-Wit" who was a young and beautiful maiden fought over by two young chiefs (two nearby volcanoes: Mt. Hood and Mt. Adams).

The volcanoes of the Cascade Mountains produce mostly volcanic ash (tephra) and pumice, rather than lava. Tephra buries older soils when it is deposited on a landscape. Many ash eruptions have occurred over thousands of years, and old soils have been buried by new layers of tephra here (Fig. 2.13) just as they have in Japan (Fig. 2.1).

When Mt. St. Helens erupted, in 1980, the top and much of the north side of the mountain were blown to pieces (Fig. 2.14). A very destructive process! Volcanic ash and rock fragments were spread for hundreds of kilometers, and all of the forests on the north and northeast side of Mt. St. Helens were killed and any preexisting soil was buried or destroyed (Fig. 2.15).

In the years since the 1980 eruption, and the destruction of the forests to the north, life has reentered the landscape and a new ecosystem has begun to develop (Fig. 2.16). The process of soil formation had to start over again in many areas after the 1980 eruption (Fig. 2.17). Young soils often contain very little nitrogen, a nutrient essential for plants. Thus many of the first plants to colonize the ash and pumice after the eruption were nitrogen-fixing plants, which means they have symbiotic bacteria in their roots that gather nitrogen from the atmosphere and share it with the plant. Even in areas where there are remnants of the old soils, numerous nitrogen-fixing plants will enter. As time passes, nitrogen accumulates in the soil along with organic matter and other nutrients. This allows other vegetation to grow that cannot fix nitrogen. As more vegetation grows, the rate of soil formation will also increase. Eventually, forests will return.

Fig. 2.12 Mt. St. Helens and surrounding forest prior to the 1980 eruption (Photo: United States Geological Service)

Fig. 2.13 Soil profile photographed in the 1970s shows an A horizon over B and C horizons that are mostly small pieces of pumice mixed with finer material. At the bottom is an A horizon formed in a soil that was buried under later eruptions

Fig. 2.14 The expanding "mushroom cloud" from an eruption of Mt. St. Helens in 1980 was visible 150 km away in Seattle

Fig. 2.15 The powerful blast from the 1980 eruption flattened the forest on the north side of the volcano. Trees lay on the ground pointing directly away from the center of the eruption (the crater of Mt. St. Helens is in the distance). It took tremendous force to flatten this forest which is several kilometers from the volcano

Fig. 2.16 Many nitrogen-fixing lupines now grow around Mt. St. Helens. Part of the crater is in the background

Fig. 2.17 Two soils near Mt. St. Helens after the 1980 eruption. *Left*: soil formed in about 30 cm of 1980 ash with an older buried soil still evident beneath. *Right*: soil formed in newly deposited pumice with little soil development. Over time an A horizon and the rich brown color evident in the soil of Fig. 2.13 will develop

Raining Mud: The 1886 Eruption of Mt. Tarawera in New Zealand

Mount Tarawera, a volcanic dome complex near Rotorua, in New Zealand, came to life on June 10th 1886 (Fig. 2.18). A rift opened along the summit of the volcano and a fire-fountain eruption of basaltic scoria commenced. The rising magma intercepted two adjacent lakes, causing an explosive steam-based eruption.

The preexisting landscape, including a famous tourist attraction, the pink and white silica terraces (Fig. 2.19), was destroyed. A gray mud rained down over the region much to the amazement (and in some cases terror) of the more distant residents who, in the days before radio, had no idea what the material was or where it had come from. Ash clouds were observed on ships up to 1000 km north of New Zealand.

Explosions from the eruption were heard hundreds of kilometers away in Auckland and Wellington, causing some people to think a war must have broken out. Several villages within about

Fig. 2.18 Painting of
the Tarawera eruption by
Ronald Cometti in 1986
based on historical
accounts (Reproduced
with permission of the
copyright holder:
R.J. Kearn. www.
tarawera.com)

6 km of the mountain were buried, and about 120 people were killed as a result of the eruption. Altogether about 2 km^3 of material was erupted from the mountain leaving a rift across the mountaintop (Fig. 2.20).

In some places the mud deposited during the Tarawera eruption was eroded into a large-scale hummocky rill pattern, possibly by water-rich outfall from the eruption or by rainfall immediately after the eruption. The surface then stabilized and became revegetated; however the hummocky erosion pattern is still visible in many areas over 100 years after the event (Fig. 2.21).

Vegetation became established and a new topsoil is forming (Fig. 2.22). Farmers added fertilizer and grass seed to hasten recovery of the productive potential of the land. Further from the source of the eruption, where only a few millimeters of tephra was deposited, it became mixed into the existing soil.

Fig. 2.19 The "white terraces" painted prior to the 1886 eruption when the terraces were destroyed, buried, or lost. Painted by Carl Kahler (Photographed with permission of Chateau Tongariro)

Fig. 2.20 The rift that formed across the top of Mt. Tarawera in the 1886 eruption. The dark reddish and black deposits are from the 1886 basaltic eruption

Fig. 2.21 The rill-eroded hummocky landscape formed in material deposited during the 1886 Tarawera eruption

Fig. 2.22 A thin new topsoil has formed in the gray tephra deposit in the century since the eruption. The buried soil that formed the ground surface prior to the 1886 eruption is evident beneath the gray tephra layer

The Taupo Pumice Eruption

Lake Taupo forms the heart of the North Island of New Zealand (Fig. 2.23). The serene, clear water lake was born of some of the largest volcanic eruptions known. The lake (Fig. 2.24) fills the gigantic collapse feature (caldera) formed after a series of eruptions. Huge volumes (>500 km³) of material have been blasted into the atmosphere and spread across the surrounding landscape with at least 28 recognized eruptions in the last 26,000 years.

The most recent major eruption from Taupo occurred about 1800 years ago (232 AD ± 10 years), which was before humans reached New Zealand. Tephra was blasted about 50 km up into the atmosphere. When the eruptive column collapsed, hot ground-hugging (pyroclastic) debris flows raced across the landscape for up to 80 km from the source leaving deposits over 100 m deep near the present-day lake. Some of the finer material would undoubtedly have stayed in the atmosphere for some months and circled the Southern Hemisphere.

The resulting Taupo pumice deposit (Fig. 2.25) mantles the entire landscape around Lake Taupo with thicker deposits ponded in valleys and thinner coatings on ridges. The pristine forest was knocked over in the blast, incinerated, and buried. Within the pumice the charcoal remains of the burnt forest are easily seen. Scientists have mapped out the angle of the fallen forest trees, and like a set of compass needles, they all point back to the source of the blast near the northeastern edge of Lake Taupo. The charcoal has been carbon-dated to help determine the date of the eruption.

For some time after the Taupo pumice eruption, the central North Island landscape would have resembled a moonscape with frequent dust and sandstorms as the winds picked up and moved material around. All the rivers and streams were choked with pumice for many years as each successive rainstorm eroded more unconsolidated material into the overloaded waterways. Even today, if you visit beaches in the North Island of New Zealand you can find Taupo pumice washed up with the tide as the pumice material is still being gradually redistributed, via the rivers, through the landscape.

Vegetation became reestablished, until covering the surface and stabilizing the new landscape. Over time organic matter built up to create a thin topsoil (Fig. 2.26). Where the pumice layer was not so thick, the plants gradually got their roots into underlying buried soils, and thus could thrive, in turn contributing organic matter to help build the new topsoil.

Fig. 2.23 Location of Lake Taupo, the source of the Taupo pumice eruption, in the North Island of New Zealand

Fig. 2.24 Lake Taupo fills the main caldera (or collapse feature) that formed following a huge rhyolite eruption 25,400 years ago that deposited tephra as far away as the Chatham Islands 870 km from New Zealand. Pumice from the most recent (232 AD) eruption is evident, washed up on the lakeshore in the foreground

Fig. 2.25 Taupo pumice exposed in a cliff. *Left*: near the eruption source, the pumice can be over 100 m deep. *Right*: carbonized logs within the Taupo pumice, 60 km from the eruption source, were forest trees that were knocked over by the explosion and mark the direction of the blast. Here the white pumice is overlain by younger, brown-colored tephras from more recent andesite eruptions from neighboring volcanoes

Fig. 2.26 Taupo pumice
soil has a topsoil
developed over weakly
weathered pumice. The
thickness of the pumice
generally decreases with
increasing distance from
the eruption source
(Photo: Glen Trewick)

Jeju, Korea: A Land of Wind and Stones

Situated at the southern end of the Korean Peninsula, Jeju is an island formed from basalt volcanoes, some of which last erupted less than 10,000 years ago. The people of Jeju describe their island as the land of wind and stones. Much of the erupted material that formed the island is highly porous scoria, as evidenced by layers of rubbly eruptive deposits and the Dolharubang ("stone grandfather" statues carved from basalt scoria) that watch over the lands (Fig. 2.27).

The resulting soils (Fig. 2.28) are stony and highly porous, such that traditional rice growing is not possible as water will not pond on the soil surface. However, the soils have been cultivated for over 1000 years to produce vegetables and citrus fruit. The stones that are a great nuisance for cultivation have, over the centuries, been collected up and formed into low walls, known as Batdam (Fig. 2.29). There are over 22,000 km of Batdam on Jeju Island. The Batdam are referred to as the "20 000 km black dragon" as they resemble a long black dragon winding through the landscape.

The Batdam are valued as an iconic feature of the Jeju landscape and an important part of the agricultural heritage of the island. Batdam perform a number of useful functions, marking field and crop boundaries (Fig. 2.30). Jeju Island is prone to strong winds as well as heavy seasonal rainfall. The Batdam provide a windbreak and prevent surface water from flowing over long distances, thus providing an important mechanism for preventing soil erosion. The windbreaks also help protect crops from wind damage. The Batdam provide fences to prevent the horses, a legacy of the Mongol arrival about 1700 years ago, and other wandering animals from damaging cultivated fields and orchards.

Fig. 2.27 The soil parent materials of Jeju Island have been erupted from predominantly basalt volcanoes. *Left*: a coastal outcrop in which the products from many eruptions are evident. *Right*: one of Jeju's many Dolharubang (stone grandfathers) carved from local basaltic scoria

Fig. 2.28 A stony soil typical of those on Jeju Island formed in the basaltic volcanic deposits

Fig. 2.29 Batdam (stone walls built from basaltic scoria boulders removed from adjacent soils) winding through the landscape of Jeju

Fig. 2.30 Cultivated fields in Jeju, Korea, surrounded by Batdam (stone walls) constructed from scoriaceous boulders gathered from the fields

Celebrating the Productivity of Volcanic Soils

When new volcanic material accumulates in thick deposits, it buries the landscape and destroys the preexisting soil and vegetation, so the process of soil formation starts again from scratch. Weathering processes begin and gradually nutrients are released, clays start to form, plants become established, and organic matter starts to accumulate, and thus a new soil develops. Basalt and andesite lava flows and tephra deposits weather to productive, nutrient-rich soils that support a wide range of crops (Fig. 2.31). In warm wet climates, weathering processes act strongly, so soils on older lava flows eventually become highly weathered to infertile clays and oxides with the nutrients leached out. Where unconsolidated tephra deposits occur, soil forms more quickly than it will on solid rock.

Where andesitic volcanic ash falls, the fine tephra allows plants to establish within a few months of the eruption and soils form relatively quickly. Andesitic tephra is relatively nutrient rich and forms soils with excellent physical properties. Where soils have been built up from many thin layers of tephra, the preexisting soil is not destroyed, and with each new eruption, material is added to form soils with excellent properties for supporting plant growth (Fig. 2.32).

Soils formed from andesitic and rhyolitic tephras often weather to form a clay mineral called allophane. Allophane makes a soil soft and friable with a low density so plant roots can easily extend deep into the soil to extract the generous amounts of water stored there. Thus plants can survive extended dry periods more readily in allophanic soils than in many other soils.

The friability of soils formed on andesitic tephra means they are easily worked to create great seedbeds and so are excellent for growing vegetables including root crops such as carrots and potatoes (Fig. 2.33). The main drawback is that allophane has tremendous capacity to hold onto phosphorus stopping plants from accessing it. Thus to maximize crop growth, it is best to add smaller amounts of phosphate fertilizer more frequently on allophane-rich soils so that the plants have a chance to absorb the fertilizer before it all becomes bound by the soil.

Fig. 2.31 Volcanic soils on basalt support a wide range of crops. *Left*: pineapples growing on an old soil formed from basalt in Puerto Rico need fertilizers to be productive. *Right*: citrus fruit growing in soils formed from basalt in Jeju, Korea

Fig. 2.32 Soil formed
from andesitic volcanic
ash. The thick, dark
topsoil is the result of a
high organic matter
content due to the great
soil conditions for plant
growth. The soil is
friable, so it is easy for
plant roots to penetrate
deep into the soil

Fig. 2.33 Productive soils formed on andesitic and basaltic tephras in Japan are friable and easily plowed to form excellent seedbeds. *Left*: planting out vegetable seedlings. *Center*: onion crops thrive. *Right*: extensive cabbage plantings on higher altitude soils formed on tephra

Soils formed on pumice and ash materials erupted from rhyolite volcanoes are often extremely low in nutrients as rhyolite is very high in silica with low quantities of other essential elements such as potassium or magnesium. Where the pumice layers are thick, it is often difficult for plants to establish. Thin pumice layers may allow plant roots to access underlying buried soils and thus grow well once established.

When people first attempted to undertake pastoral farming on the pumice soils in the central North Island of New Zealand, the sheep and cattle became ill and died from a deficiency disease that became known as "bush sickness." Pine trees were found to grow well as their roots penetrate deeply, going through the pumice to the underlying buried soils, and they did not suffer nutrient deficiency problems. Thus large areas of New Zealand pumice soils were planted into the forest (Fig. 2.34).

The mystery of "bush sickness" was unraveled by scientists who found that the bush sickness was caused by a deficiency of cobalt which can be remedied by adding cobalt to fertilizer or giving stock vitamin B12. With a growing demand for dairy products, some of the plantation forest areas are being converted to dairy farms once the trees have been harvested.

Fig. 2.34 The Kaingaroa Forest—the largest plantation forest in the Southern Hemisphere—was planted in soils formed on rhyolite pumice in the 1930s. The soils were then considered useless for pastoral farming as animals became ill with "bush sickness"

Chapter 3
Soils Born of Water

Tending rice in water-saturated soils in Thailand

© Springer International Publishing Switzerland 2016
M.R. Balks, D. Zabowski, *Celebrating Soil*, DOI 10.1007/978-3-319-32684-9_3

Introduction

"Water is the driving force of all nature"

Leonardo da Vinci

Water is vital to all life on Earth and is also essential to soils and the organisms that live within them. Water erodes soil and deposits material from which new soils form. Water is key to many physical and chemical weathering processes that convert rock materials to productive soil. The amount of water that a soil can hold, and make available to plants, influences the kinds of plants and animals that can survive in any given environment, thus underpinning the terrestrial ecology of our planet. Soil water maintains the dynamic interactions between roots and soil, providing nutrients to plants.

From catastrophic floods to subtle leaching of wetland soil, water acts to change soils. River floodplain landscapes contend with periodic inundations from flood waters, but the soils get fresh deposits of new parent materials. Saturated soils undergo chemical changes that give them a unique set of soil properties (Fig. 3.1). The accumulated remains of plants form peat (or organic) soils in environments where soil surface saturation prevents decomposition of plant materials.

Just as excessive water creates unique soil properties, so too can too little water. Dry soils may not support many plants and very little organic matter accumulates. Salts may accumulate in dry soils and impact on the ability of the soil to support plant and animal life. Salt and soil acidity are issues in soils on coastal margins where seawater has had an influence.

Fig. 3.1 Soil formed in an area with an extremely high rainfall (about 5000 mm/year). The *blue-gray* colors are indicative of a soil where there is no free oxygen due to water saturation (Photo: New Zealand Society of Soil Science)

Water-Saturated Soils

Under saturated conditions, all the oxygen in the soil atmosphere can be consumed by microbes (many of which, like humans, take in oxygen and respire carbon dioxide), leaving the soil in an anaerobic (no free oxygen) state. A lack of oxygen leads to changes in soil chemistry and biology, and the resultant soils are often referred to as "gley" soils.

Iron oxide minerals change from being bright orange (rust colored) and insoluble when they are oxidized, to being colorless and soluble when they are in anaerobic (low oxygen or reducing) conditions. Manganese oxide minerals also change in response to the presence or absence of oxygen. Soils that are saturated for long periods thus lose their warm brownish colors, leaving them pale gray or sometimes blue-gray (Fig. 3.2).

Many soils have a seasonally fluctuating water table (depth at which the soil is saturated) and are saturated in some seasons and drier in others. As a soil dries out and the water table drops, air moves into the pores in the soil that were previously filled with water, allowing oxygen to interact with the soil. Where air penetrates into a soil, possibly down old root channels or cracks, the iron and manganese minerals that were dissolved in the water under anaerobic conditions react with the oxygen and become insoluble, precipitating out as rusty brown iron oxides and black manganese oxides. The tell-tale sign of oxidation-reduction processes occurring in a soil is a pattern of bright orange spots and blackish specks or concretions on a pale gray background (Fig. 3.2). If the soil is always wet, it may all be gray or bluish in color.

Fig. 3.2 Soil with a high water table. The lower part of the profile is a *bluish gray* color showing that it is permanently saturated with water. The mid-part of the soil has a pale-colored matrix with some rust-colored iron oxides and black manganese oxides present in areas, such as old root channels, where oxygen can diffuse into the soil

Fig. 3.3 Iron oxide minerals naturally precipitate out in drains or stream beds that carry iron-rich water from anaerobic groundwaters. Here the iron oxides are visible as the rust-colored material on the stream bed

Sometimes in streams where water is draining from saturated soils, the water has a bright orange/brown color, or you will see a rusty brown-colored "slime" accumulating on the stream bed (Fig. 3.3). The rusty brown material is simply iron oxide (ferrihydrite) compounds that have formed and precipitated out when the dissolved iron from the anaerobic drainage waters comes in contact with oxygen in the stream water. There are specialized bacteria that gain energy from oxidizing the iron minerals and form the slime found in these environments. The bacteria can appear as an "oily-looking" sheen on the reddish oxides. Precipitation of iron oxides can sometimes cause problems by blocking pipes that are carrying groundwater from an anaerobic environment.

In saturated conditions, the microbial population becomes dominated by organisms that survive by extracting oxygen from compounds such as nitrates and giving off by-products including nitrogen gas, carbon dioxide, and methane. Oxidation of methane, as it escapes into the atmosphere, may contribute to the 'will-o'-the-wisp' effect of ephemeral flames sometimes observed on wetland and swampy areas.

Saturated soils often lose much of their coherence, thus making them difficult to walk over as you sink into them. The combination of water saturation and lack of coherence also makes them difficult soils to sample (Fig. 3.4).

Fig. 3.4 Sampling a saturated soil can be a messy business. Here an auger was used to extract soil samples which were laid out on a board. This soil is rich in clay and organic matter. Beneath the surface, the soil is anaerobic (it has no free oxygen) and has a dark gray-blue color. However once it is exposed to air, the soil rapidly changes to the brownish oxidized color of the soil surface

Soils Formed on River Floodplains

Soils on floodplains build up as the periodic inundation of flood waters deposits silt, sand, gravel, and other materials (Fig. 3.5). Over time deep soils form as each flood deposits new material that buries the previous soil, building up many layers over time. Given enough time, a new topsoil forms after each flood as plants reestablish at the new soil surface.

Soils formed on lowland river floodplains are some of the most productive soils. Hence human agriculture and settlements are often situated on floodplains. The river provides a water source and often also easy transport to and from the area (Fig. 3.6). The disadvantage, of course, is that such land is prone to inundation by floodwaters.

Where rivers drain glaciers, or highly erodible rapidly uplifting mountain chains, they carry huge sediment loads that fill and block the river channel causing the river to overflow and diverge into new paths. The river thus migrates back and forth across the floodplain, building braided river channels and depositing new layers of gravel, sand, and silt (Fig. 3.7).

The soils on river floodplains vary depending on the sediment load of the river, the frequency and size of floods, and the height of the water table which is related to the height of the floodplain above river level (Fig. 3.8). Where rivers are fast flowing, they often carry a high sediment load of larger materials (boulders, gravel, and sand) which are deposited to form rapidly draining, low-nutrient

Fig. 3.5 River floodplains are subject to periodic inundation. *Left*: Floodwaters overflow a river stopbank to inundate an area of farmland in New Zealand. Here the river stopbank is designed to be overtopped so the excess flood waters safely flow out into an area that is free of houses, thus protecting a town further downstream from flooding. *Right*: mud (silt and clay) deposited by a major flood can be seen covering the floodplain of the Waipaoa River in New Zealand. The river waters, though having now receded into the river channel, are still a pale color due to the load of mud they are carrying

Fig. 3.6 Soils formed on the floodplain of the Urubamba River in Peru are important to the local people. The flat fertile soils, with an adjacent source of water, are the most intensively farmed and productive in the area

Fig. 3.7 Alluvial plains near Golmud in China. *Left*: braided river channel of a tributary of the Yangtze River that drains water from the rapidly uplifting and eroding margin of the Tibet Plateau. *Right*: the vast alluvial plains have formed at altitudes of over 4000 m and are frozen in winter. Because of the inhospitable climate and short time since deposition, the gravelly sand-dominated soils are weakly weathered with relatively low nutrient availability and low organic matter content. The naturally sparse vegetation is heavily grazed in summer

Fig 3.8 Soils formed on river floodplains. *Left*: a soil, formed in silt deposited within the previous decade, has only a shallow, weakly developed topsoil due to the short time since the last inundation of silt-laden flood water. The incipient topsoil is, in turn, likely to be buried by the next flood event. *Right*: soil formed in gravelly sand deposited over 12,000 years ago, during the waning phases of the last glaciation, by a sediment-loaded braided river

Fig. 3.9 A river floodplain. *Left*: the levee or raised bank that has formed on this floodplain is shown up by the pattern of the fences. *Right*: the same area (though photographed from a somewhat different angle) a few days after a flood event with water still lying in the backswamp areas

soils. In lowland areas where the river moves more slowly, with a low gradient, the coarse sediments are left behind and only fine clays and silts are deposited. Where the water table is near the ground surface or where fine materials prevent rapid drainage of water through the soil, swampy conditions prevail and soils are saturated with water for long periods.

Natural levees form adjacent to streams on river floodplains (Fig. 3.9). Levees are raised banks that run parallel to the stream. When a river floods, as the water overtops the riverbank, the water flow slows down, and much of the coarser suspended sand and gravel material immediately settles out of the water forming a raised bank. The water that flows further away from the river carries only finer silt and clay material which may take days to settle out of the water. Once water overtops the levee, it cannot easily flow back to the river when the water level recedes. A swampy area called a "back-swamp" forms where the water is trapped until it evaporates or soaks into the ground, leaving all its sediment load behind. Thus soils on river floodplains often have a pattern of coarser sandy soils near the river and finer clays in the backswamp areas behind the levees.

Many river floodplains are now protected by artificial flood "stop banks" or "dykes" which are built to constrain flood water in the river and thus protect the adjacent houses, cities, and farmland from inundation (Fig. 3.10). While flood stopbanks prevent much damage and heartache from floods, there are a number of adverse consequences of stopbanks. As the stopbank prevents water, and therefore also the sediments and accompanying nutrients, from spreading onto the land, the sediments are carried to the sea, speeding up the silt accumulation in estuaries and river deltas. Other effects are the buildup of silt within the riverbed, resulting in the need for ever higher stopbanks and an increased danger to floodplains should the river ever breach the stopbank. The lack of regular inputs of sediment to the soils mean that no new nutrients are being added, thus fertilizers are required to maintain soil fertility.

Fig. 3.10 Flood stopbanks protect farmland and other infrastructures on the floodplain of the Waipaoa River in New Zealand. Sheep graze the unprotected area inside the stopbank but can be readily moved if flooding is imminent. The protected area outside the flood stopbank is used for more intensive development such as cropping, vineyards, and housing

Water Scarcity: Soils and Salt

Salts accumulate in soils mainly as products of rock weathering and soil formation. In wetter regions, water flows into the moist soil and gradually seeps down to the groundwater, carrying dissolved salts with it. Ultimately the groundwater seeps, via springs, into streams and rivers and hence to the oceans. The dissolved salt the water carries to the sea has gradually accumulated in the oceans, and thus the sea is salty. However in drier regions, where the potential for water evaporation is greater than the rainfall, soils are rarely wetted enough for water to flow through the soil.

Where there is little, if any, water seeping from the soil to the groundwater salts gradually accumulate in the soil. Salt-affected soils occur in more than 100 countries worldwide and may cause serious damage to a soil's productive potential. Areas of salt accumulation often occur naturally in low-lying areas where occasional runoff events carry water and dissolved salts. When the water evaporates, the salts are left behind at the soil surface (Fig. 3.11). Types of salts vary but common varieties include sodium, calcium, or magnesium chlorides and carbonates.

Soils with high salt contents often have pHs as high as 8 or 9 and are inhospitable environments for all but the most salt-tolerant plants (Fig. 3.12). In some cases, rare and interesting plants and other organisms occur that are adapted to the salty environment. The salts may accumulate as discrete layers often at the soil surface. Where some rainfall occurs, salts may be leached (moved in solution) partway into the soil and a salt-rich layer may occur at depth within the soil.

Fig. 3.11 Surface waters in Australia. *Left*: in many parts of semiarid Australia, once every few years, there is a heavy downpour and the resulting flood waters spread across the landscape. *Right*: the flood waters dissolve salt from the soil surface then flow to low-lying areas and where the water evaporates, leaving the salt behind, thus removing the salt from some areas and concentrating it in others

Fig. 3.12 Profiles of salt-affected soils from Otago, New Zealand. *Left*: a soil where salt accumulation precludes all but the most salt-tolerant vegetation. *Right*: the white layer in the subsoil is calcium carbonate, which in this semiarid environment is leached partway down the soil profile by occasional rainfall events, but there is insufficient moisture to leach it from the entire profile (Photos: New Zealand Society of Soil Science)

Fig. 3.13 Diagram showing relationship between water table and surface salt accumulation

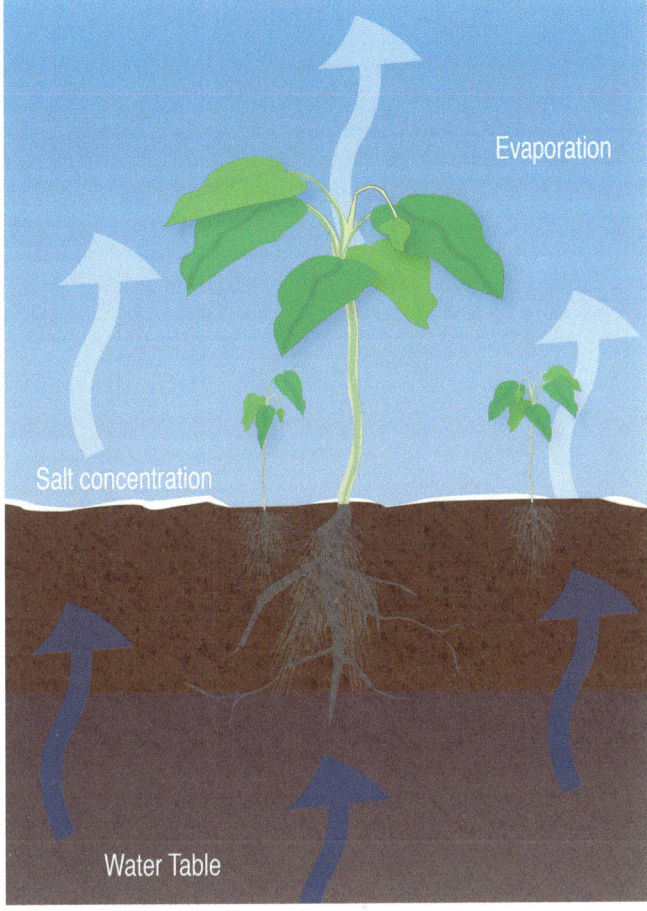

Salt accumulation in soils can be exacerbated as a result of human activities that lead to a rising soil water table including irrigation, urbanization in upstream areas, and tree removal to facilitate agricultural activities such as cropping. In Australia a condition known as dry-land salinity occurs in farmland in semiarid regions where trees have been removed and so are no longer removing water from deep within the ground leading to a gradual rise in the water table and accompanying salts.

Accelerated salt accumulation in soils occurs where human activities (most often excessive irrigation without adequate drainage) cause the water table to rise until it gets within a meter or so of the soil surface. Once the water table is near the land surface, moisture is drawn to the surface and evaporates, leaving the salts behind in the soil (Figs. 3.13 and 3.14). The problem is made worse when the groundwater, or the water used for irrigation, is rich in salts.

Over the centuries many people in dry regions have suffered as a result of crop failure due to salt accumulation in soil. Mesopotamia (the region of the floodplains of the Tigris and Euphrates rivers—now in Iraq and Iran) is often referred to as the "cradle of civilization." Irrigation was first developed in the Mesopotamia region about 8000 years ago, and the resulting productivity enabled the first advanced human society to flourish for thousands of years. However, a lack of drainage, which could have ensured the salts were washed through the soil and removed, led to salt accumulation. Salinization of the soils, silting up of canals, and periodic floods that destroyed infrastructure and caused changes to river courses, were ultimately major contributing factors in the decline of the ancient civilization and the development of the harsh desert environment in the region that has persisted for the last 600 years.

Fig. 3.14 A flagpole floating in a shallow well is an indicator of water table height and thus potential for salt damage to soil. If the *red* part of the flag is showing above the soil surface, then the water table is within a meter of the surface and thus there is danger of salt accumulation on the ground surface. Recent irrigation has left the ground surface wet, but the poor vegetation cover indicates that this area has a high salt content. Shepparton, Australia

Much of Australia is arid or semiarid with potential for salt accumulation in soils. About one third of Australia has sodium salts in the soils, and about 16 % of Australian farmland is estimated to be impacted by salinity that has been aggravated by human activities. Some consider salinity to be Australia's most significant environmental issue. Salt accumulation in the soil leads to a decline in soil productivity. One example of such decline comes from the Carnarvon area in Western Australia.

For much of the year, the Gascoyne River near Carnarvon in Western Australia is dry (Fig. 3.15). Farmers extract water from the ground beneath the river bed to irrigate their extensive banana, and other fruit, crops. Periodic flood events refill the groundwater below the riverbed. A flood in 2010 passed through a particularly salty area and recharged the shallow groundwater reservoir with salty water. As a result, the water in the shallow groundwater under the riverbed, which the Carnarvon farmers depended on for irrigation, became too salty to use and the local fruit-growing industry suffered. A new flood has now recharged the groundwater and flushed the salts out of the system and thus allowed the shallow groundwater in the river bed to again be utilized.

Fig. 3.15 The Gascoyne River in Western Australia. *Top left*: farmers extract irrigation water from the shallow ground-water beneath the often dry riverbed. The signs (*top right*) and the drought-stressed banana plantation (*bottom*) tell the story of the effects of salt in the groundwater, rendering it unsuitable for use for irrigation in 2013

Frozen Water: Glacial Deposits

Glaciers have covered much of the Northern Hemisphere many times in the last 2.6 million years during repeated glaciations. Many of the older glaciations left deposits that were removed by more recent glaciers. However there is still abundant evidence of the last glaciation. The last glaciation ended by about 12,000 years ago leaving behind extensive areas of highly variable glacially deposited material that forms a variety of soils. We will look at three of the most common glacial deposits: till, outwash, and glacial lake deposits. Till is formed when glaciers grind rocks into smaller pieces which are ultimately left behind when the glacier melts. Till can form from rocks, gravels, sands, and silts that are all mixed together, ground up, and compressed by the weight of ice pressing down on them. The loose rocks, gravels, and sands carried on or in the ice are often deposited on top of the compacted till when the glacier melts (Fig. 3.16).

As an ice sheet moves, particularly when it is melting back, the meltwater forms streams and rivers flowing off the ice. The flowing water is called glacial outwash and often carries a large amount of ground-up rocks, gravel, and sand from within the glacier. The outwash deposits are the rock materials that are carried and then deposited, by the flowing meltwaters (Fig. 3.17). The bigger rocks are deposited by fast moving flood waters. Smaller gravels and sands are deposited, often in layers, where the water moves more slowly. As rock materials are carried in rivers, they are ground into finer particles. The finest silt particles were often blown off the river floodplains onto surrounding land to form a deposit known as loess.

Glacial streams often flow into glacial lakes where extensive deposits of silt and clay form as the fine particles settle out in the still water. Lake sediments often have annual deposition layers formed

Fig. 3.16 A coastal
bluff, in Northwest
United States, showing a
variety of glacial
deposits. Some till is at
the top with glacial lake
sediments and glacial
outwash deposits below

during changing conditions throughout the year. Spring melt brings in a layer of coarser particles and ice on the lake surface in winter creates still conditions for a layer of fine clay to settle out (Fig. 3.17).

Soils that form in various glacial deposits have varying rock content and texture and therefore a variety of soil properties, particularly related to drainage and aeration (Fig. 3.18). Soils formed in glacial lake sediments are fine textured, retain water, and can easily become waterlogged. The saturated soils may be gray or have mottles of red and gray colors in the upper soil profile and be all gray and gleyed in the lower soil profile.

Glacial outwash deposits are usually well drained. In some cases, there may be finer-textured layers interbedded among coarser layers of outwash. The fine layers help retain some water in the soil by restricting downward water flow from the fine-textured material into the underlying sandy-rocky layer until the finer-textured soil is saturated with water.

Soils formed in till are more variable than those formed in outwash or lake sediments as the type of till or the combination of till types can change the soil properties. In some cases, where the till is entirely loose material from melting ice, the soil can be well drained and well aerated with yellow-brown colors. When compacted till is present, it may be so compacted that almost no water will flow through it. Soils on compacted till can be very-poorly developed as roots cannot grow into it and compacted till can be so hard it is almost impossible to dig, even with a pick.

Fig. 3.17 Glacial deposits. *Left*: outwash deposits with layers of gravel and stony materials interbedded with sandy layers. *Right*: glacial lake sediments showing annual deposition layers (known as varves). There is also a small amount of gravelly till mixed with outwash above the lake sediments

Fig. 3.18 Soil profiles developed in three types of glacial deposits. *Left*: a forest growing on an excessively drained outwash soil (Photo: Stan Gessel). Center: a very wet, poorly drained soil developed in glacial lake sediments. *Right*: a soil formed in loose till over compacted till. The upper horizons of this soil are well drained and aerated, but the compacted till (just at the bottom of the profile) restricts water flow in the lower soil making it waterlogged in winter and spring

Peatland Soils

Swamps, mires, bogs, and wetlands all are areas of land that are saturated with water (Fig. 3.19). When plants die in waterlogged areas, the remains of the plants lie in the water. Any oxygen is quickly used up by microbes leaving anaerobic conditions where decomposition of organic matter is often extremely slow. Thus the dead and decaying plant matter slowly accumulates, forming peat (Fig. 3.20). Over thousands of years, the peat may slowly accumulate to form a deposit many meters thick. The plants that survive on the surface are adapted to growing in an environment with plentiful water but low soil oxygen and often low nutrients. Sedges, rushes, mosses, and plants that can send oxygen down to their roots are common in peatlands. The organic soils that form in peatlands vary from clearly recognizable remains of leaves, branches, and roots to strongly decomposed black humic material which feels as smooth as silk when you rub it between your fingers.

Peats, because they accumulate slowly over time, contain records of past environments. Scientists study the pollen and plant remains in peat soils to understand past changes in vegetation which can give clues to past climates and human activities in the area. Other materials may be preserved in peats such as layers of tephra that record the timing and extent of past volcanic eruptions and charcoal which provides a record of past fires. Peat soils are not strongly coherent so are generally not suitable to build houses or other structures on without special engineering efforts (Fig. 3.21).

Peat soils accumulate slowly as the organic matter gradually builds up. However peat can be rapidly destroyed. Fires, once started, in peat are difficult to control and may burn for years. Peat in many

Fig. 3.19 The wetland margins of small lakes, such as this one in Western Australia, are areas where water-tolerant plants grow—gradually extending out and infilling the lake with plant material that ultimately forms peat soils

Fig. 3.20 Organic soils formed in peat. *Left*: a peat soil in a forest in the USA. The light band near the bottom is lake sediment composed of diatoms, tiny siliceous shelled organisms that live in lakes. The dark band above is soluble organic matter that has migrated down and accumulated above the diatom layer. *Right*: a closeup view of peat from a sedge/scrubland peat bog where some of the remains of plant materials are clearly identifiable

places, such as Ireland, is mined as a source of fuel or potting mix. If peat is drained, the surface will naturally subside, first as the water is removed and then as the newly aerated material decomposes. Organic materials such as peat decompose to carbon dioxide and water; thus, a solid soil is transformed into components of the atmosphere or flows away as water.

Some of the largest areas of peat soils are those in the arctic. Much of the arctic peat is underlain by permafrost, ground that is "permanently" frozen (Fig. 3.22). There is concern that continued warming could lead to melting of the permafrost and subsequent decomposition of the peat, potentially releasing both carbon dioxide and methane (which are greenhouse gases) into the atmosphere.

Fig. 3.21 Keeping power poles upright on peat soils is challenging. Even with extra support, the lack of coherence in the underlying organic soil material is evident to all

Fig. 3.22 Arctic peat soils on the North Slope of Alaska. *Left*: while the near surface soil is thawed in summer and can be cut with a spade, the deeper soil is frozen solid, so a corer was used to extract it. The white-colored parts of the core are ice. *Right*: the North Slope of Alaska, near the Arctic Ocean, is a vast swampy tundra plain. There is a high water table and the area is interspersed with many small lakes in summer and covered in snow in winter

Soils Formed in Lake Sediments

Sediments accumulate on lake beds. If a lake drains, when a barrier is breached or a dam is removed, the sediments are exposed and soils form. Many rivers have been dammed to provide hydroelectric power or store water (Fig. 3.23). However, over time as sediment accumulates, the volume of water that can be held in the dam's reservoir decreases. Dams also change rivers and may increase or decrease erosion downstream from a dam and impact on riparian ecosystems and river fisheries. At some locations dams are being removed from rivers and new soils begin to form in the sediments left behind. The removal of two dams on the Elwha River in the Olympic National Park in Western Washington State, USA, provides an example of the fresh sediments left behind when a dam is removed and the river returns to its former course (Fig. 3.24). The two dams were originally built to

Fig. 3.23 A dam in the Swiss Alps. The pearly color of the water is due to the refraction of light by the high silt content of the water. Some of the silt will settle out and accumulate as sediment in the bottom of the dam

Fig. 3.24 The Elwha River in Western Washington State after dams were removed. *Left*: Site during removal of one of the dams in about 2012. *Right*. View of some of the sediments remaining after the dam was removed. An area of nearly 3 km^2 of new lake sediments was left after the dams were removed

Fig. 3.25 Sediments left behind following removal of dams on the Elwha River in about 2012 and examples of the soils that are likely to form on the sediments over time. *Top left*: coarse sediment comprising sand, gravel, and boulders, deposited near the river inflow to the dam. *Top right*: a soil formed on similar coarse sediments from the glacial lake that drained over 12,000 years ago. The *brown color* shows that oxygen is present in the soil as air can move through the

provide electricity and water. However the dams also stopped a large salmon run in the river. A decision was made to remove the dams and return the river to a more natural state, reestablish the salmon run, and reforest the sediments deposited behind the dams.

In the past, glaciers carved the Elwha River Valley. Glacial rivers in this valley were dammed by a large ice mass at the mouth of the valley. Thus a large lake is formed that left sediment deposits just as would form behind any other dam. The lake sediments were exposed when the last glaciation ended about 12,000 years ago and the ice dam melted. The soils formed in the glacial lake sediments are, therefore, now over 12,000 years old and give us an opportunity to see how soils in the new dam sediments might develop over time (Fig. 3.25).

Coastal Margins: Acid Sulfate Soils, as Sour as Lemon Juice

In many parts of the world, often due to population pressure, humans now occupy areas on low-lying coastal margins. The soils in areas previously inundated with seawater on coastal margins tend to have high water tables and there are challenges to sustainably manage them. In their natural state, the sediments, often mud (silt and clay) or sand, are saturated with water and thus anaerobic (Fig. 3.26). Many

Fig. 3.26 A tidal river in northern New Zealand with mangrove mudflats on the margins (*left*) and a saturated tidal mudflat soil with mangrove roots (*right*)—the brown color of the oxidized soil surface gives way to dark blue-gray oxygen depleted mud beneath

---

Fig. 3.25 (continued) large pores between the coarse particles. There is an O horizon of forest litter at the soil surface as a coniferous forest is established here. *Bottom left*: fine silt and clay sediments, deposited nearer the dam, have cracked as they dried and will be easily eroded until new vegetation cover is established. *Bottom right*: soil formed in silt and clay sediments from the glacial lake. The *pale gray color* shows a lack of oxygen penetrating much of the soil as there are few pores that air can move through. Like the soil above this, soil formed over 12,000 years ago and is under forest. The differences between the two soils are mainly due to the difference in sediment size

Fig. 3.27 A coastal margin area south of Sydney, Australia, that has been drained and developed for farming. *Left*: pastoral grasslands developed using irrigation, drainage, and inputs of lime to combat low pH. *Center*: an area where the soil pH has dropped to about three and plants are unable to survive. *Right*: a profile of the drained soil. The *yellow* lenses are a sulfate mineral called jarosite which forms in the low pH environment

coastal wetlands are important habitats for a unique range of plant, bird, and fish species including mangroves and salt grass-dominated ecosystems.

Where areas that were formerly tidal mudflats are drained, the sulfur present in the sediments may be oxidized to form sulfuric acid and the soil pH may drop to levels as low as 2—similar to lemon juice (Fig. 3.27). Where the pH drops below about four metals such as arsenic and aluminum become soluble with concentrations that may be toxic to plants. The acidic drainage waters can cause fish kills and other ecological damages and are corrosive to concrete and steel structures such as sewage or water pipes and bridge piers.

Celebrating the Productivity of Water-Influenced Soils

Rice is the staple food for about half of the world's population and has been cultivated in China for at least 7000 years. In some countries in Southeast Asia, rice provides up to 80 % of human food. Over 100 countries produce rice and most is consumed by the local populations in the regions where rice is grown. The labor-intensive practice of transplanting rice (Fig. 3.28) has been conducted for at least 1800 years. Wild rice originally grew in wetland areas. Cultivated rice has relied on monsoon rains to provide the water-saturated soils needed to grow most varieties. However, people have been manipulating water tables for thousands of years to ensure rice growing areas have adequate water.

The warm climate and saturated soil environment allow rice to produce more food energy per unit land area than any other grain crop, an important feature as we work to feed the increasing world population. Since around 1950, increases in fertilizer use and improved rice varieties and cultivation practices have enabled rice production to keep pace with population increases. The challenge is to maintain production into the future.

The soils that result from periodic flooding and addition of silts and sands form river floodplains that are flat and fertile, thus creating some of our most productive soils. In the Nile River Valley in Egypt, the annual flood was vital to soil productivity as it deposited nutrient-rich black silt and water onto the floodplain, thus ensuring a successful harvest. The height of the annual Nile flooding was critically important, too high, and houses and roads were damaged, too low, and crops would fail for lack of water. The annual flooding of the Nile is still celebrated today by a holiday in Egypt. The floodplain of the Nile sustainably supported a population of about two million people for thousands of years. Life is more challenging in Egypt now as the population has grown to about 84 million. Since 1970 floods have been controlled and prevented by the Aswan High Dam and fertilizers and irrigation are required to sustain soil productivity.

Fig. 3.28 Planting rice and growing rice plants in Thailand

The Huang He or Yellow River in China is one of the great rivers of the world—it rises in the mountains of Tibet and flows in a big loop through the erosion-prone loess plateau then east across China to the sea south of Beijing. Often referred to as the "cradle of Chinese civilization," the Huang He is sometimes depicted as the river mother (Fig. 3.29) as it nurtures the people of China. The English translation of the river's name, "Yellow River," refers to the yellow color of the water that is the result of the extraordinarily high sediment load, which the river carries from the loess plateau.

The average sediment load carried to the sea by the Huang He over the last 1000 years has been estimated at about 1 billion tons per year (Fig. 3.30). Much sediment is also deposited in the lower reaches of the river. Over thousands of years the flood-deposited sediments have built up rich yellow alluvial soils on the 400,000 km^2 North China Plain forming one of China's most important agricultural areas. The soils support corn, sorghum, millet, wheat, vegetables, and cotton.

Fig. 3.29 Statue on the
banks of the Huang He
at Lanzhou depicting the
Huang He as the River
Mother

Fig. 3.30 The sediment-laden water is evident as tourists, on a traditional sheep-skin raft, navigate the Huang He (Yellow River) at Lanzhou in China. Eroded dry hills in the headwaters of the Huang He provide a source of sediment

Inevitably, larger floods cause the river to break its banks and flood extensive areas of land, often resulting in devastating damage and millions of deaths. Hence the Huang He is also often referred to as "China's sorrow." About every hundred years or so, there have been major changes to the river's course with the mouth of the river sometimes moving by as much as 400 km along the coast.

Fig. 3.31 A vineyard established on the floodplain of the Waipaoa River in New Zealand is protected by a flood stop bank. Twenty-five years earlier, a major cyclone deposited over 1 m of silt across this land surface

Water from the Huang He supports about 140 million people and is used to irrigate about 74,000 km^2 of land. Since the 1950s, there has been a 60 % decline in the flow in the Huang He, attributed to a big increase in irrigation and human consumption as well as a decline in precipitation in the region. Huge efforts at soil erosion prevention on the loess plateau have led to a decline in sediment load of the Huang He from about 1.2 billion tons per year between the 1920s and 1960s to about 200 million tons per year between 1990 and 2009. The decline in sediment load, along with the decreased discharge, has led to huge changes in the sediment regime in the river delta. From about the year 2000 the delta stopped extending outward and began to erode.

Soils on river floodplains, once protected from frequent flooding, are often highly productive and resilient (Fig. 3.31). It is often possible to reestablish crops on flood alluvium which brings new nutrients into the soil. The underlying buried soil may also be reached by plant roots, and thus may also provide support for crop growth.

Many crops are grown on organic soils, ranging from pasture to potatoes. However, there are some plants, such as blueberries, that are particularly well adapted to the peatland environment (Fig. 3.32). Thus, soils formed under the influence of water, whether they are on river floodplains, managed by flooding for rice growing, or formed from organic peat materials are a vital resource for human food production.

Fig. 3.32 Blueberries growing on peat soils in New Zealand ripen around Christmas time making a niche crop for export to the Northern Hemisphere winter

Chapter 4
Soils of Forests

Western larch growing in shallow, rocky soil in the Cascade Mountains of Western USA

© Springer International Publishing Switzerland 2016
M.R. Balks, D. Zabowski, *Celebrating Soil*, DOI 10.1007/978-3-319-32684-9_4

Introduction

"A nation that destroys its soils destroys itself. Forests are the lungs of our land, purifying the air and giving fresh strength to our people."

Franklin D. Roosevelt

Some of the world's most extensive landscapes are forested. From the broad boreal forests of the high northern latitudes to dense rainforests of the tropics, forests can be found on almost all types of landscapes as long as there is enough precipitation and adequate soil to support their growth (Figs. 4.1 and 4.2). Environmental factors such as moisture, temperature, and topography will determine exactly what type of forest and soil occurs on a landscape and helps create variety in both. One consistent feature of forest soils is that a tremendous amount of litter from leaves, bark, and branches is dropped from trees to the surface of the soil every year. Regardless of forest type, forested landscapes can be magnificent and have some amazing soils.

Fig. 4.1 Young soil under a conifer forest

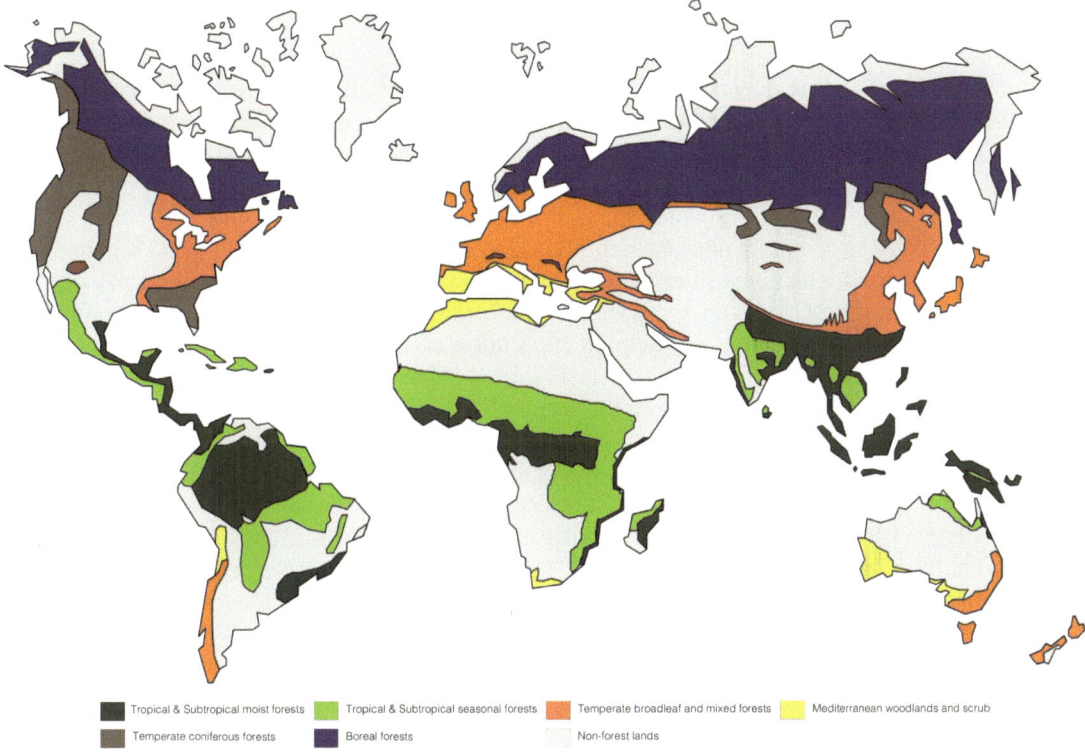

Fig. 4.2 Approximate major forest biomes of the world. Non-forested land includes grasslands, wetlands, deserts, tundra, and bare land

The variability among forests and trees is incredible. Giant sequoias and redwoods can have diameters of nearly 10 m and reach heights of over 100 m. Yet there are pygmy forests, where stunted trees may only reach a few meters tall—a mere fraction of their normal height. Pygmy forests usually occur when the soils supporting these trees are shallow and not very fertile or where a cold climate limits growth. Forests can also vary in density and understory vegetation. If availability of water is limiting, forests will have trees widely spaced allowing a lot of light to reach the ground. Forests with a canopy cover so thick that very little light reaches the ground have few understory plants and a cooler microclimate under their canopy. Such differences in environmental conditions affect litter decomposition and development of soil horizons. Disturbance is another factor that affects life in forests. Disturbance opens patches within forests allowing new seedlings and understory plants to begin new, young forests among the old that eventually grow into mature forests. Disturbances will also change the soil and can create pockets of different soil types in areas that might otherwise seem uniform.

Because forests occur on so many different landscapes with such a wide range of climates, they can have extremely diverse soils. A riparian forest may grow along a river on a young soil developed in alluvial deposits with abundant cobbles and sand. Old soils that are almost all clay can support many types of temperate and tropical forests. Forests located in warmer and drier climates will not only look very different from those in a moist, cool area, but the fundamental processes that develop their soils may be quite different.

Temperate Broadleaf and Mixed Forests

Temperate broadleaf forests may be either deciduous or evergreen. Broadleaf forests typically occur where there is ample moisture available in summer to support an extensive canopy of foliage and usually have some soil moisture available throughout the year. The canopy of deciduous forests can be very dense in summer blocking light to the ground, often resulting in understory species blooming in spring while more light is still available. Soils of deciduous forests can vary with the rock type and soil age, but usually have a more nutrient-rich A horizon and O horizon, although the O horizon may be thin. Because deciduous forest litter tends to have a higher nutrient content, it generally decomposes faster keeping the O horizon somewhat thinner and letting nutrients from the decomposing organic matter leach into the A horizon helping the A horizon stay fertile.

There are many varieties of broadleaf forests. Figures 4.3 and 4.4 show mixed broadleaf deciduous forests (also called hardwood forests) from the eastern USA and from near the Great Lakes region of the USA. Both soils under these forests are very old with well-developed A and B horizons and abundant clay. The O horizons will become thick during the fall when leaf litter accumulates but decrease in depth as decomposition reduces the amount of organic matter present until the next autumn when fresh leaf litter is added again. Snowpacks can also compress the litter, and snowmelt provides moisture to aid in litter decomposition in the spring. The cycling of nutrients released from decomposing litter, combined with weathering and atmospheric deposition, helps maintain soil fertility. Maples, oaks, chestnuts, hickory, walnut, and many other hardwood species are found in the Great Lakes

Fig. 4.3 A stream surrounded by mixed hardwood forest from the eastern USA with a nearby soil profile. The soil developed in ancient sedimentary rocks and is very old. About 2 m down are some remnants of sedimentary rock structure beneath heavy clay

Fig. 4.4 A mixed deciduous forest and soil from the Great Lakes region, USA. The soil is clayey and has mottles of color in the lower profile. The high clay content can restrict water and oxygen movement particularly at depth in the soil, forming mottles of *gray* and *orange* colors

region. Obviously, since they are called hardwoods, these trees have hardwood in their boles, and many are highly valued as wood for furniture and flooring.

There are many types of broadleaf forests that are not mixed and are dominated by a single type of tree, such as eucalyptus or beech forests (Figs. 4.5 and 4.6). Eucalyptus forests are most common in Australia, although plantations of eucalyptus are now found in many other countries where they grow rapidly. Most eucalyptus trees are evergreen and some are among the tallest trees on Earth. One type of mountain eucalyptus grows to over 90 m tall. Eucalyptus often grows on nutrient-poor soils. These trees can have deep roots that are able to access water from very deep in the soil. Many species of eucalyptus are adapted to fire and resprout after a fire if the root systems are not killed.

Another type of broadleaf forest dominated by a single type of tree is the southern beech. Southern beech forests are widespread in the Southern Hemisphere from southern New Zealand and Australia to southern South America (true beeches are more common to mixed deciduous forests of Europe and North America).

Fig. 4.5 Eucalyptus growing in New South Wales, Australia, on an old, deep soil. This soil has shallow O and A horizons. Iron and aluminum oxides are abundant, and the iron oxides give the soil its yellow-orange color

Fig. 4.6 A southern beech from New Zealand with a clay-rich soil that previously supported a beech forest

Fig. 4.7 An aspen forest near the Rocky Mountains in the Western USA

Aspen is another predominantly single-species broadleaf deciduous forest (Fig. 4.7). Aspens are found in the Northern Hemisphere, across North America, Europe, and Asia, often at higher latitudes or elevations with less developed soil. Aspen forests are adapted to fire and can resprout from their root systems following a burn.

Temperate Coniferous Forests

Temperate coniferous forests occur where there are seasonal droughts that these forests can tolerate. Water may be limited due to summer droughts, or there may be limited water available throughout the year. Coniferous forests can retain their foliage for many years and usually have needles that are waxy and limit moisture loss through transpiration. Coniferous forests will drop some older foliage after the needles (or scales) are shaded or when the needles are so old they no longer photosynthesize effectively. Some coniferous trees are the tallest known trees, and some are the most massive. Old-growth temperate coniferous forests can be hundreds to thousands of years old; such coniferous forests develop an amazing cathedral-like environment beneath their canopy.

Fig. 4.8 Two rotations of slash pine growing on a sandy soil in southeastern USA

Changing coniferous forest type and soil forming factors such as climate causes formation of different types of forest soils. Although coniferous forest soils typically have a dry season, they can have varying amounts of rain or snow throughout the rest of the year. These soils can be young or old and exhibit horizons that show intense leaching and poorly developed horizons or show deep accumulations of organic matter in an O horizon. The soil types are as variable as the environments and parent materials in which the soils form! Soils get unique additional characteristics from the type of forest they are under. For example, some forests have much more acidic soils than others, and some have much more organic matter in the litter layer. Often the litter of coniferous forests has a lower nutrient content and will decompose more slowly, producing organic acids. Some coniferous forests are also adapted to frequent fires which can destroy O horizons, increase soil pH, and alter soil development.

Pine (*Pinus*) forests (Figs. 4.8, 4.9, 4.10, and 4.11) are widespread throughout the Northern Hemisphere. They are often found in drier areas with coarser-textured soils that drain well although

Fig. 4.9 Radiata pine growing on an outwash soil in southern New Zealand. Radiata pine is not native to New Zealand but grows very fast there; trees are harvested after about 25 years of growth

some tolerate wetter soils. There are many varieties of pines from small shrubby trees, to rapidly growing flatland pines, to tall, stately ponderosa pine trees (growing over 80 m tall). One of the world's oldest individual organisms is the bristle cone pine (some are over 4000 years old).

Another coniferous tree common in the Western USA and Canada is Douglas-fir (Fig. 4.12). This species can reach hundreds of years in age. It is also a valuable commercial species. Douglas-fir can grow in drier or wetter locations and tolerates various soil types, but prefers a well-drained soil. Douglas-fir is a good example of a coniferous tree that can grow in either young or old soils and on a variety of parent material types and topography.

Some coniferous forests such as true fir (*Abies*) forests can accumulate substantial amounts of litter in O horizons that are acidic and lower in nutrients. In wetter environments, organic acids can leach out of O horizons into the mineral soil causing intensive weathering in the surface mineral horizon

Fig. 4.10 A mature
ponderosa pine from
Western USA

creating a podzolic soil (Fig. 4.13). Iron, aluminum, and soluble organics leach (eluviate) out of the
mineral soil forming an E horizon. The iron, aluminum, and organic acids move down and accumulate
in an illuvial B horizon. Although podzolic soils are acidic and not the most fertile soils, they can be
quite beautiful with their bright contrasting colors. Certain understory plants (e.g., heathers or blueberries) also produce acidic litter that encourages this eluvation and illuviation.

Some old coniferous forests regenerate best following disturbance. Disturbance can be from fire,
windstorms, windthrow of individual trees, diseases such as root rots, insect infestation, or humans.
Anything that opens up a forest allowing more light to reach the soil can encourage young trees to
begin growing fast. Thus many old-growth forests have soils with mixed horizons from trees pulling
up soil when they fall and mounding it. Old-growth forests typically have a mixture of older and
younger trees with an irregular micro-topography from such disturbance. In some cases, disturbance
can kill most of the forest (such as a high-intensity wildfire). Some disturbances such as landslides not
only remove the forest but also the soil (Fig. 4.14).

Fig. 4.11 Mature Scots pine in front of a younger pine forest in Scotland

Fig. 4.12 An old-growth Douglas-fir forest with a soil profile developed in glacial outwash and till from the Pacific Northwest, USA. This soil has an orange B horizon showing an accumulation of iron and aluminum oxides

Fig. 4.13 A subalpine forest of the Cascade Mountains of Washington State, USA, with a soil profile that is characterized by eluviation and illuviation. Abies and ericaceous species are common in this type of forest and their litter promotes the formation of acidic soils

Fig. 4.14 Soil erosion after severe wildfire in a subalpine forest. Note the thick ash covering the soil surface

Boreal Forests

Boreal forests (also called taiga) cover a huge area of the Northern Hemisphere and are mostly found in Russia, northern Europe/Scandinavia, Canada, and the USA/Alaska (Fig. 4.2). Globally, boreal forests are the most extensive type of forest.

Boreal forests typically begin about 50–60° latitude north and continue north until the climate becomes too harsh for trees to grow. In some places, boreal forests start at more southerly latitudes, such as Central Canada, where the continental climate is particularly cold. Along coastal areas the latitude where boreal forests begin is often farther north as the maritime climate is not as harsh as the continental interior. Precipitation is variable for boreal forests; however, most precipitation occurs as snow that melts in the spring in time for early growth; there is usually some rainfall during summer months.

There are a variety of tree species found in boreal forests, but the most common are pines, spruce, birch, aspen, and larches. All are trees that can withstand long harsh winters and short summers. The growing season can be very short (3–4 months), so growth rates are usually slow. Although boreal forests are typically fairly dense, trees do not normally grow extremely tall or large. Understory vegetation is often limited due to the density of the forest canopy, but mosses, lichens, and *Vaccinium* species such as blueberries are common (Figs. 4.15 and 4.16).

Fig. 4.15 A Canadian boreal forest from Ontario with a soil profile that shows eluviated (E) and illuviated (B) horizons. Mosses and lichens cover most of the forest floor

Fig. 4.16 Russian taiga with a deep soil that shows an E and illuviated B horizon. The B horizon of this soil is less subalpine forest of the Cascade Mountains reddish showing less accumulation of oxides. Extensive lichens and moss are present on the forest floor

Soils under boreal forests are typically acidic, have thick O horizons, and are podzolic, with iron, aluminum, and some organic matter moving into the B horizon from an E horizon in the upper mineral soil (Figs. 4.15 and 4.16). This occurs when the soils are well drained. However, many boreal soils can be very wet due to a lack of drainage caused by water restriction from shallow bedrock, dense glacial deposits, permafrost, or impermeable soil. Such sites can form swampy forests and wet, organic-rich soils (Figs. 4.17 and 4.18).

Wildfires are critical for succession in boreal forests. Vast areas of boreal forest burn every year, but that does not mean these forests die. Usually, even with severe fires, pockets of intact forest remain and reseed surrounding soils. Jack pines are widespread in Canadian boreal forest and have serotinous cones that open up to release seeds following fire. Jack pine can reseed rapidly after fire. Black spruce will rapidly reestablish in wetter boreal soils after fires, whereas Jack pine will reestablish faster in areas with drier soils. Aspen will resprout if its roots are not killed. Nutrients that were bound within the organic matter of the thick O horizons are released in ash after a fire and become available for new seedlings that grow in the open areas (Fig. 4.19).

Fig. 4.17 Russian boreal forests with permafrost at depth. *Left*: a boreal forest near the Pinega River floodplain. *Right*: a boreal forest surrounding a small lake

Fig. 4.18 A wet floodplain boreal forest and soil in central Alaska. This soil has permafrost at depth; ice lenses are present in the mottled B horizon. Mottling in this soil occurs because the permafrost prevents water drainage. Soil at the base of the profile is frozen showing ice lenses. Each *yellow* and *black bar* indicates 10 cm

Fig. 4.19 Two Canadian boreal forests soon after wildfire. Wildfire has not killed all the trees in the forest. *Right*: a regenerating Canadian boreal forest several years after a wildfire. The extensive woody debris on the forest floor will decompose slowly to become an integral part of the O horizon

Tropical and Subtropical Forests

When most people imagine a dense tropical or subtropical forest landscape and its soil, they picture a very dense rainforest with a soil that is very old and deep with an orange-red color. This is the classic image of a tropical forest with a type of soil often called laterite (Figs. 4.20 and 4.21). Tropical environments that are hot and have abundant precipitation allow trees to grow rapidly—vegetation is dense with a tremendous variety of species. If the landscape is also old and the parent material has abundant iron and aluminum, then the high rainfall and warm temperatures will help these soils develop a deep, red, lateritic soil profile. To form lateritic soils, intense leaching removes silica and most weatherable minerals rapidly; some clays and iron and aluminum oxides will form. The oxides will persist in the soil even when the soil is hundreds of thousands of years old. Although lateritic soils may not be very acidic, they usually have lower fertility, partially due to the lower concentration of organic matter in the soil and the leaching loss of nutrients as the soil developed.

Most of the nutrients in a lateritic soil are in the organic matter and living plants. With the moist environment and warm temperatures, litter will decompose rapidly releasing nutrients quickly. Although these soils may support a productive forest that is dropping a lot of litter to the O horizon, the O horizon will be shallow and few nutrients will move into the mineral soil because the forest will rapidly take up any nutrients released (Fig. 4.22). Clearing forests with lateritic soil can lower fertility if organic matter is lost, requiring the addition of nutrients to grow crops. Some cleared lateritic soils

Fig. 4.20 A deep, highly weathered, old tropical forest soil in Thailand. This soil is rich in iron and aluminum oxides. Weatherable minerals have been broken down and soluble nutrients have leached. The iron oxides give this soil its characteristic bright orange-red color

Fig. 4.21 A Peruvian tropical rainforest near the headwaters of the Amazon River with old soils. The deep, red, lateritic soils have abundant iron and aluminum oxides and clay. The high clay content means unpaved roads can easily trap vehicles

Fig. 4.22 A lateritic soil in southern Brazil that was under forest and now cleared for crops

can irreversibly dry and harden making them difficult to manage for farming. With slash-and-burn agriculture, forests are typically cut down, the debris left from harvesting is burned, and then crops are planted. Using slash-and-burn agriculture, the soil will temporarily gain nutrients released by burning, but both the loss of organic matter and nutrient removal with harvesting will deplete soil nutrients in a few years requiring people to abandon these sites and move to another area to begin the cycle again. Forests will return to abandoned slash-and-burn areas, and if enough years pass before people return to burn the forest and grow crops again, the soils can recover.

Many tropical and subtropical forest soils are not lateritic (Figs. 4.23 and 4.24). A tremendous variety of soil types occur beneath these forests, from poorly developed sandy rocky soils to older soils that are high in clay but not considered to be a laterite because they are not dominated by iron and aluminum oxides that formed within the soil. Tropical soils can be young, on parent materials that are not high in iron, at locations that are not well drained, or disturbed too often to allow development into laterite. Non-lateritic tropical soils are often more productive than lateritic soils as they have more nutrients giving them better fertility.

Fig. 4.23 A subtropical forest growing on limestone that is part of a sinkhole filled with water in Mexico. Such soils can be very productive and accumulate substantial organic matter in a dark calcium-rich A horizon

Fig. 4.24 A young subtropical alluvial soil that is on a floodplain in Puerto Rico. This soil was under forest but is now used for pasture

Mediterranean Woodlands and Savannas

Mediterranean woodlands and savannas are characterized by a hot, dry summer season with limited moisture in the soil for plant growth. These are not dense forests; trees are typically widely spaced to minimize competition between trees for water, particularly with savannas. Mediterranean woodlands may have a wetter, somewhat cooler winter although forests may still have limited water availability. Mediterranean forests are not widespread, but do occur in a variety of locations on all continents (Fig. 4.2). Broadleaf trees are common, particularly with the oaks of southern Europe and southwestern USA, and Eucalyptus forests of southwestern Australia. Pines and evergreen trees can occur in sparse forests in northern Africa. Savannas are drier woodlands with more widely-spaced trees. Wildfires, and human-caused fires, are a common disturbance to woodlands and savannas so many species are adapted to survive fire (as long as fires are neither too severe or too frequent) either by having thicker bark, resprouting from roots, or producing abundant seed. However, when drier forests are near dense human populations, the frequent lower-severity fires are often suppressed to protect lives and property leading to a buildup of plants and litter which are fuel for larger more catastrophic fires. When fires occur in areas with such a fuel accumulation, fire severity can be high resulting in devastating impacts. Wildfires and wide spacing of trees that drop litter can result in little O horizon accumulation in these forests.

The wide variety in the type of soils that occur under Mediterranean forests and savannas shouldn't be surprising as these forests are widely scattered across all continents (Figs. 4.25 and 4.26). Soils range from old with clay and oxide-rich horizons to young soils with weakly developed horizons. A drier climate which limits moisture to plants and less plant litter are common factors in soil

Fig. 4.25 An oak forest from southern Oregon growing on an older soil with a clay-rich B horizon and dark A horizon

Fig. 4.26 An African savanna with baobab trees from Lake Manyara National Park. There are smaller trees between the baobab intermixed with grasses. The small trees are acacias, a shrubby tree that has nitrogen-fixing bacteria that will add nitrogen, an important nutrient to the soil. These trees are growing on a very old soil that is rich in silt and clay. The soils are very dry half of the year, and wet the other half (Photo by Dale Cole)

development. Limited moisture means that a well-developed soil under these forests is usually very old if the soil has a high clay content. The clay can retain soil moisture but still dries out in summer.

Some Special Forest Trees

The world's largest and tallest trees are redwoods. These trees are native to the west coast of the USA (particularly southern Oregon and northern California) (Figs. 4.27 and 4.28). The prevalent coastal fog helps provide moisture to these trees even during the dry summers allowing them to have a long growing season and reach a massive size. You might think that the soil supporting such trees would be exceptional, but these soils are not particularly different from other forest soils, typically having a simple profile of O, A, and B horizons.

Kauri are large coniferous trees native to areas of northern New Zealand (Figs. 4.29 and 4.30). Kauri have beautiful wood and many were harvested in the 1800s and early 1900s, often for use as ship masts. Large kauri are now scarce. Large kauri developed podzolic soils underneath. Kauri drop extensive litter; water flowing through the litter and down the bole encourages leaching of the upper mineral soil forming an E horizon that is deeper near the trunk of the tree and becomes thinner away from the bole (Fig. 4.31).

Fig. 4.27 An extensive canopy soil of organic matter can accumulate in redwoods and support a variety of epiphytic plants (*above*). Above *right*: deep accumulations of litter build up at the base of redwoods when the soil is not disturbed

Fig. 4.28 A large redwood tree growing near the border of California and Oregon in Western USA and the soil that supports this redwood showing O, A, and B horizons

Fig. 4.29 Large quantities of epiphytes grow in the canopy of kauri

Fig. 4.30 Visitors viewing the largest living kauri (possibly 2000 years old), named Tane Mahutu by the Maori. The Maori consider these large trees sacred

Fig. 4.31 Podzolic soils that developed under kauri show intense soil leaching and development of an E horizon. *Left*: the E horizon is thicker where the soil was closer to the bole of the original tree. *Right*: a soil that was formed under kauri and now supports an exotic pine plantation shows mixing in the upper E horizon from harvesting and development of a new type of O horizon with pine needles

There are many other special forest types and trees with unique characteristics or soils. We've included a few more examples here with some interesting features (Figs. 4.32, 4.33, and 4.34).

Fig. 4.32 Western red cedars are another large, long-lived North American tree that can grow on a variety of soil types but prefers a moister environment and soil. Therefore, cedars are widespread along the rainy Pacific Northwest coast of North America. The bark of these trees is stringy and was stripped off by North American natives and used for many purposes such as in making baskets and rope and in weaving. Cedar is very rot resistant and the wood was also used for canoes and in construction of long houses

Fig. 4.33 Sitka spruce also grows tall and large in the Pacific Northwest. This spruce prefers a moist environment and soil. Wood from spruce trees is very light and was formerly used to make airplanes

Fig. 4.34 Mangroves grow along coastal areas within salt-water tidal zones. Mangrove seeds float in water and eventually grow when they wash up on a suitable shore—thus mangroves are found around the world. *Right*: mangroves have specialized roots that grow above the water saturated soil and are thought to assist with both support and oxygen uptake

Celebrating the Productivity of Forest Soils

Probably the first thing that people think about when considering productivity and forest soils is the production of wood. Certainly, wood production, whether it is for creating lumber or paper, is the most widespread use of plantation forests. Plantation forests are forests that are specially managed to provide wood products over multiple harvesting rotations. Plantation forests are found throughout the world and differ from natural forests because there is often only one species of tree and is valuable. Many coniferous or softwood trees are commonly grown in plantations because they are valuable, grow faster, and are easier to harvest (Fig. 4.35). Tree species that are native to one area, but grow particularly well in another area, are often used in exotic plantations. For example, eucalyptus is native to Australia but is grown in many parts of South America. Likewise, Radiata pine is native to Western USA but is widely grown in Australia and New Zealand (Fig. 4.36).

Intensive, managed forests can produce the wood products that people need without damage to soils if the plantations are managed wisely. Forest soils have a wide range in productivity, but when trees are harvested, nutrients are removed, soils can be damaged by compaction from heavy equipment, and harvesting on steep slopes can cause erosion. Best management practices vary from place to place, but in general, harvesting on flat or gently sloped ground, protecting mineral soil by keeping organic horizons intact over mineral soils, minimizing the use of heavy equipment, and leaving branches and foliage (which have a higher nutrient content) on site are good practices. Nutrients removed with harvested wood may need to be replaced to maintain productivity (Fig. 4.37).

Forests provide us with much more than just wood products for building homes and other structures. They provide valuable wood for furniture, paper, and crafts. One of the most widespread uses of wood globally is as fuel for cooking fires. We also get many highly-valued foods from forests including a variety of edible vegetation, nuts, berries, and mushrooms. Floral arrangements often include forest foliage. Many wildlife species need forest habitats to survive. Another critical function

Fig. 4.35 Harvesting of southern pines in the southeastern USA

Fig. 4.36 A plantation of Radiata pine in New Zealand with a variety of stand ages from harvesting at different times

Fig. 4.37 Helicopter fertilizing forests in Western USA. *Right*: a slice from a Douglas fir bole that shows annual growth rings. Note the increase in the width of the growth rings that occurred after fertilization

of forests is that forest soils can filter and clean water providing freshwater to lakes, rivers, and groundwater, ultimately giving us accessible sources of clean drinking water. Many municipal water supplies originate in forested watersheds.

Ecosystem services is a term that is often used to encompass the many essential things that we get from forests and their soils that are not easily valued but are vital to our well-being. Water, habitat, recreation, and aesthetics are all examples of priceless commodities that are not easy to assign a dollar value. Nevertheless, many forested areas are key providers of ecosystem services (Figs. 4.38, 4.39, 4.40, and 4.41). Often the soils that support forests are not as fertile as grassland soils, and clearing forested soils for agricultural is likely to provide soil that is not very good for agriculture and elimi-nate many of these ecosystem services. Clearing forests has often occurred because forested lands have not been valued, and economically people could benefit more in the short term from other land uses. Nevertheless, forested lands have great intrinsic value, and we need to protect forested lands and their soils to maintain the ecosystem services we often take for granted.

Fig. 4.38 Mountains, glaciers, and forests are part of a watershed providing water to a lake in Glacier National Park, Montana, USA

Fig. 4.39 Deer and giraffe are two examples of wildlife that need forest habitat (giraffe photo by Dale Cole)

Fig. 4.40 Morel mushrooms (*upper*) and huckleberries (*lower*) are two examples of food we get from forests

Fig. 4.41 Wood is often used as fuel for cooking fires and heating, especially in developing countries

Chapter 5
Soils of Grasslands and Rangelands

Ripening corn in a New Zealand field

© Springer International Publishing Switzerland 2016
M.R. Balks, D. Zabowski, *Celebrating Soil*, DOI 10.1007/978-3-319-32684-9_5

Introduction

"To be a successful farmer one must first know the nature of the soil."

Xenophon, 400 BC

The broad temperate plains of the Earth support vast expanses of magnificent grasslands and range-lands. Whether you call such areas steppes, prairies, chaparral, or simply grasslands and rangelands, these landscapes have some of the Earth's most fertile soils (Fig. 5.1). Thus, many grasslands and rangelands are now used for farming and grazing.

Water is usually the most limiting factor to the growth of plants in grasslands and rangelands. Often there is an extensive dry season. Rangelands are normally drier than grasslands. The shortage

Fig. 5.1 A grassland soil from the Western United States with a deep A horizon that extends to 70 cm. The A horizon of this soil is rich in humus, but does not look very dark because the soil is dry

of water limits the type and amount of vegetation; thus, drought-tolerant grasses and shrubs dominate rather than trees. Grasslands and rangelands support two dominant types of plants: ones that grow fast and flower as soon as the soil warms when there is still plenty of water available but die when the soil dries, or grasses and shrubs with long roots that can access water from deep beneath the soil surface after the dry season has begun. Periodic droughts often affect these areas and can last for years. Nevertheless, when water is available in spring, the landscapes can be covered with blooming flowers and abound with wildlife.

Why are grassland soils so fertile? Some grasslands have the deepest and best A horizons of any soils on the planet with abundant humus and excellent physical properties. The best soils under grasslands or rangelands have deep, nutrient-rich topsoils with little leaching of nutrients. The humus in grassland soils comes from the extensive fine root systems of the grasses that continually grow, die, and thus add large amounts of organic matter directly into the mineral soil as they decompose and become humus. Grassland ecosystems store carbon in a reverse arrangement to forests. Where forests have most carbon above ground and less in the soil, most of the carbon stored in grasslands is below ground in the soil. From the landscape perspective, it may seem as if a grassland ecosystem doesn't have much carbon compared to a forest, but grassland and rangeland soils hold so much organic matter they can have almost as much carbon as a tall, mature forest ecosystem.

All of the continents except Antarctica have areas of grasslands and rangelands. Such prairies and steppes tend to have gentle landscapes with immense vistas. They can be some of our most beautiful environments and soils.

Loess: Some Soils Start with Wind

Loess is an accumulation of fine wind-borne deposits of particles that are predominantly silt but can include substantial sand and clay (Fig. 5.2). Loess is a parent material often found under grassland and rangeland soils (Fig. 5.3). Loess can form large hills similar to sand dunes or form flat broad plains with gentle topography (Figs. 5.4, 5.5, and 5.6). Loess deposits can be hundreds of meters deep; the depth of the deposit and the texture of the soil will vary with finer-textured soils found farther away from the source of the Eolian material.

Loess can accumulate in an area where there is a source of fine particles such as a dry lake bed or broad areas of glacial outwash; frequent winds lift and carry the material many kilometers toward an area of less wind allowing the particles to settle. Many of the greatest loess regions of the world were formed during past glaciations. When sheets of ice covered vast areas of the planet, there was extensive erosion from ice with rocks grinding against each other, creating many fine particles that were carried downwind of glaciated areas. The extensive ice sheets created high winds in areas near the glacial margins due to the changes in air temperature between glaciated and non-glaciated areas. The high winds would pick up the fine particles and blanket huge areas of the landscape near the glacial margins, burying the old topography in broad plains or creating rounded hills similar to giant sand dunes.

Soils formed in loess usually have a silt-dominated texture that is relatively consistent with depth, a deep A horizon if there is adequate rain to support grasses, and clusters of particles that form columnar aggregates. Columnar aggregates are evident in Fig. 5.2 where the linear cracks show the space between aggregates. The silt texture and aggregation usually allow roots to penetrate deeply, although occasionally a dense B horizon will form. Insects and arthropods are also abundant, as are many mammals that form extensive burrows. Loess soils are generally easy to dig, yet animal burrows can retain their form and do not collapse. In some locations, where the loess hills are large, there can be differences in soils between the north and south side of a hill due to differences in temperature, moisture, and erosion. For example, there may be fewer grasses and more shrubs on a northern slope in the

Fig. 5.2 A silt-rich soil formed from loess in the USA. The soil has a dark A horizon and extensive aggregation with particles grouped into columnar-shaped aggregates

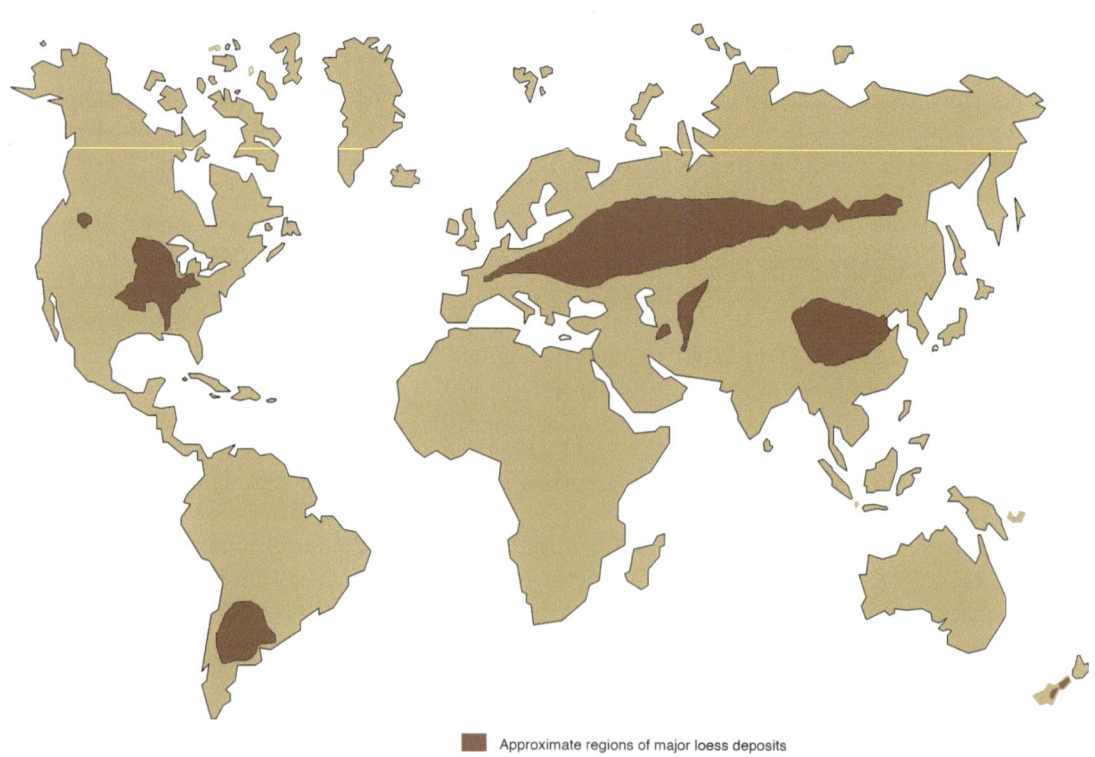

Approximate regions of major loess deposits

Fig. 5.3 Approximate regions of major loess deposits. Many smaller areas of loess can be found across Asia, Africa, and Australia

Fig. 5.4 A rounded hill in Northwestern United States formed from loess deposited during past ice ages

Fig. 5.5 Winter wheat growing on a soil developed in loess in the Western United States. Note the gentle topography of the rounded hills

Fig. 5.6 Pasture on a loess soil in Poland

Northern Hemisphere. With fewer grasses, soils will have a shallower A horizon because fewer grass roots means less organic matter enters the soil directly from decaying roots; thus less humus forms making the A horizon shallower.

Grasslands

Because grasslands are so productive and easily adapted to agriculture, few grasslands have native vegetation. For example, one of the most endangered ecosystems in North America is native prairie. Most native grasslands have been converted to agricultural use, typically for crops but some for grazing. The rich dark topsoils (Fig. 5.7) indicated their fertility to European settlers and much of the original grasslands were plowed under. Because moisture availability is normally the biggest limitation to plant growth, these soils are either irrigated to produce crops or crops are grown that can overwinter and mature in spring while there is still moisture in the soil. Winter wheat is an example of a crop grown over winter for early harvest in spring (Fig. 5.7).

Although many of the original grasslands of North America developed on soils formed from loess, not all grassland soils form in loess. There are many excellent grasslands that have developed on floodplains (Fig. 5.8). Broad valley floodplains where slow-moving river water deposited fine silts and clays can develop deep A horizons and become excellent agricultural soil. Grasslands with soils developed on limestone can have a particularly dark A horizon and are called rendzinas (Fig. 5.9).

Fig. 5.7 A field of winter wheat growing on a loess soil of Washington State, USA, with a very deep organic-rich A horizon. The B horizon of this soil is a pale yellow-brown from the formation of some iron oxides and clay. Note that the deepest horizon (beginning at 130 cm) is a white color—this is from an accumulation of calcium carbonate that has weathered out of the loess and leached down the soil to this depth. Soils with a higher rainfall would have all of the carbonates leached, but the limited rainfall here moves carbonates down but not out of the soil

Fig. 5.8 A floodplain soil of the Willamette Valley, Oregon, USA, that is rich in clay and farmed. This soil has been ripped to break up a plow pan

Fig. 5.9 A shallow
rendzina soil formed
over limestone, South
Island, New Zealand

Grasslands: Central China

Classic deep loess grassland soils are found in a vast area of Central China (Fig. 5.10). These are rich soils and plant productivity is mostly limited by the availability of water. Although loess soils have been used for crop production for thousands of years in China, they have remained productive largely due to the application of organic wastes to the fields. The additions of plant, animal, and human wastes have helped add nutrients back to the soil and helped bind particles together into aggregates to reduce erosion (Fig. 5.11). Loss of nutrients and erosion are two problems associated with loess soils used for agriculture, and adding organic matter helps prevent both. Wheat is one crop grown in Central China (Figs. 5.12 and 5.13).

Fig. 5.10 A deep loess soil of Central China. This soil can be 4–5 m deep

Fig. 5.11 An aggregate taken from the soil shown in Fig. 5.10

Fig. 5.12 Separating
wheat germ from chaff

Fig. 5.13 Straw produced from the soil in Fig. 5.10

Grasslands: Australian Wheat Belt

The Australian wheat belt stretches across a 13,000,000 ha arc roughly parallel to the east coast of Australia from Queensland through New South Wales and Victoria to South Australia and up the west coast of Australia to just north of Perth. The Australian wheat belt is mainly confined to the area with winter- or spring-dominated rainfall that totals at least 400–600 mm/year. The wheat belt could also be called "productive grasslands" as they are also used for grazing (Fig. 5.14). Merino sheep provide meat and wool, but as the returns are less than for wheat, sheep farming tends to be concentrated at the less productive margins of the wheat belt.

Australia is one of the world's largest producers and exporter of wheat, supplying about 15% of the global export market, thus making an important contribution to global food resources. The soils across the vast Australian wheat belt have a lot in common: they are generally all in semiarid environments on ancient, strongly weathered surfaces (Fig. 5.15).

Soils range from sands to clays and have a very low organic matter content, a result of the hot climate with limited rainfall and regular plowing, so any plant material that reaches the soil rapidly decomposes. Where the soil parent materials are windblown sands, likely deposited within the last 100,000 years, such as those near the west coast north of Perth, the soils are often deep and well drained and have yellower colors compared to the red colors of many of the highly weathered soils of the areas further inland (Fig. 5.16).

Fig. 5.14 Australian productive grasslands. Merino Sheep graze in the foreground. Wheat is the green crop in the middle with yellow canola flowers visible in the middle distance

Fig. 5.15 Soils of the Australian wheat belt. *Left*: a typical soil lacks a dark-colored topsoil due to the low organic matter content. *Right*: a prepared and sown wheat field awaits rain to support germination

Fig. 5.16 *Left*: sandy wheat belt soil in Western Australia. *Right*: a cemented horizon from an inland wheat belt soil. Iron-rich pebbles and coarse quartz sand grains are visible. The cemented layer can form a barrier to root penetration and thus limit soil productivity

Grasslands: Alpine Meadows

Alpine meadows have stunning scenery and can have interesting productive soils. However, alpine environments have short growing seasons—snows come early, winter is long, and spring and summer are short. At higher elevations, the climate becomes too cold and plants will not grow which limits soil formation.

In alpine meadows, grasses and herbaceous plants are most common, although some locations are dominated by shrubs or a mixture of plants (Figs. 5.17 and 5.18). Many alpine soils are shallow due to erosion or very slow weathering rates. Soil development is usually slow in such cold environments,

Fig. 5.17 Alpine meadows with long-term human use for livestock grazing in Switzerland's Alps

Fig. 5.18 Alpine meadows in the Cascade mountains of Washington State, USA. This site is a wilderness area with limited human impacts

Fig. 5.19 An alpine soil (*left*) from a meadow on Mt. Rainier in USA (*center*). The darker A horizon is about 20 cm deep probably due to mixing from burrowing mammals such as pika (*above right*) and marmots (*below right*). Evidence of disturbance can be seen where the patchy darker A horizon has been mixed into the lighter B horizon below

but some soils are deeper due to loess or volcanic ash accumulation. Many burrowing animals help to "till" the soil, and such mixing can encourage deeper A horizon formation than you might expect in such a cold environment with limited vegetation (Fig. 5.19). Although the growing season is short, alpine meadows add organic matter to soil from their litter, roots, and crusts of plants such as lichens.

Grasslands: Mounds of Mystery

Scattered across numerous prairie locations of the USA are "mounds of mystery" (Fig. 5.20). These mounds, often called Mima mounds, can be less than 1 m high to several meters tall and three to four times as wide. One of the key characteristics that make the mounds mysterious is that they typically have an A horizon that is very deep at the center of the mound and gradually tapers to become so thin at the edge of each mound that it can essentially disappear (Fig. 5.21). Similar mounds are also found in South Africa.

No one is sure how these mounds formed, but there are plenty of theories ranging from earthquakes, Eolian formations, glacial outwash deposits, to burrowing animals. Burrowing animals, such as gophers, are currently a favored theory; however, there is no direct evidence, such as bones, that indicate gophers forming such large mounds. There is indirect soil evidence that burrowing animals have inhabited these sites. Figure 5.21 shows an old backfilled burrow still evident in the C horizon of a mound that is backfilled with A horizon soil. Such round channels backfilled with soil of another horizon are typically from burrowing animals but does not guarantee that the whole mounds were formed by burrowing animals. For now, the formation of these mounds remains a mystery.

Fig. 5.20 Mounds of the Mima prairie in Washington State, USA, covered with grass and flowering camas. Camas has an edible bulb and these prairies were often burned by Native Americans to prevent forest from encroaching, so the camas would continue to grow and could be harvested annually. This frequent burning has left extensive fine particles of charcoal in the A horizon making it very dark

Fig. 5.21 A cross section of a Mima mound showing the deep A horizon in the center tapering off at the edge of the mound (*left*). This form suggests mounding of the soil occurred after the A horizon was developed. *Right*: a backfilled burrow channel believed to have formed by ancient gophers is shown below the A and B horizons in this profile

Rangelands

Rangeland landscapes are characterized by limited vegetation due to very limited availability of soil water—a combination of low rainfall and high evaporation (Fig. 5.22). Water scarcity affects soils in rangelands in a variety of ways. Typically there is less organic matter in rangeland soils since there are fewer plants producing less plant matter which means less litter that can become soil organic matter compared to moister soils. Less water flow also keeps the pH and salt content of these soils higher. Because water is so limited, the plants that live on rangelands often have roots that extend many meters deep. Deep rooting allows plants to access water that may only be available far below the soil surface. As with alpine areas, the growing season can be short, but in this case, the growing season is short because of the limited time when water is available. A brief period in spring when rains come starts plant growth, but growth quickly ends when temperatures get so hot that the soil dries out and plants can't access water.

Limited water availability also means that there is limited leaching in rangeland soils. There is not enough water to move all of the ions in solution out of the soil, so salts may move deeper in the soil profile but not out of it. Thus when weathering occurs, soluble rocks such as carbonates that easily dissolve into solution will not leach out of the soil. The calcium and carbonate dissolve into solution when water is present but may move no deeper than the depth the rain penetrates into the soil. If the average depth of rain penetration remains the same year after year or there is a barrier to water movement, then the calcium and carbonate in solution can accumulate in a subsurface horizon that is often found in rangeland soils (Fig. 5.23). Sometimes carbonate-rich horizons can become cemented and hard.

Fig. 5.22 A shallow rangeland and soil from the Western United States receiving less than 300 mm of precipitation per year. Each *yellow* and *black bar* on the tape is 10 cm

Fig. 5.23 A deep rangeland soil from a site that also gets limited rainfall. The pale horizons at approximately 70 and 110 cm deep are rich in carbonates. This soil formed in loess. The carbonates are from the dissolution of minerals in the loess parent material and are slowly moving down the soil profile where they precipitate. *Right*: Carbonates can also be found as accumulations on the underside of rocks in soil where the carbonates precipitate on the rock face as the water evaporates

Rangelands: The Pilbara Region of Western Australia

The Pilbara region of Western Australia is an arid tropical area of over 500,000 km². In summer temperatures in the open are often over 45 °C and sometimes over 50 °C. Rainfall comes in infrequent large floods with occasional tropical cyclones. The underlying bedrock is some of the oldest on the planet, with some rocks about 3.5 billion years old. A rich mosaic of vegetation occurs across the landscape (Fig. 5.24) reflecting the underlying soils and moisture availability as well as time since the last fire event.

Where the soils form in iron-rich parent materials, a dark red color prevails (Figs. 5.24, 5.25, and 5.26). On the lower-lying areas, soils range from very sandy to clay rich, but at all sites vegetation growth is limited by drought and low soil fertility.

In the hot, dry climate, soils are prone to being left devoid of vegetation, as a result of fire or grazing pressure; thus, wind and water erosion remove fine material, often forming a desert pavement (Figs. 5.25 and 5.26).

On hilltops, erosion often leaves rocks exposed near the surface, with only shallow soils forming which are, in turn, more challenging for vegetation to survive on and thus prone to further erosion (Fig. 5.27).

Fig. 5.24 *Left*: the view southwards from Mt. Bruce in the Karijini National Park shows the vegetation pattern with an iron-ore mine in the distance. Note the different shrub and tree species along ephemeral stream channels. *Right*: a ground level view of the savannah-type grasslands in Karijini National Park

Fig. 5.25 A typical soil of the Pilbara region containing iron-rich rock material. The stones concentrated at the soil surface form a desert pavement that protects the underlying material from further wind and water erosion. The distinctive stone layer near the top of the subsoil is likely to be a former surface pavement that has been buried by a later soil building event

Fig. 5.26 Rangeland in the Pilbara region of Australia. *Top left*: a reserve area with limited introduced grazing animals. *Top right*: an adjacent area of the reserve where fire has recently passed through. *Bottom left*: Cattle graze the sparse grasslands, removing more palatable species of plants. *Bottom right*: the stone-encrusted desert pavement that forms on the soil surface where plants are absent and erosion (especially wind erosion) removes finer soil materials

Fig. 5.27 Shallow stony soil formed on sandstone

Celebrating the Productivity of Grasslands and Rangelands

Our richest agricultural lands are often located on soils that once supported native grasslands and rangelands. In Europe and Asia, most of these native ecosystems were converted to agricultural lands long ago. In the Americas, Australia, and parts of Africa, agricultural development is more recent. Without the extremely productive grassland soils, we would not be able to feed the world's population. The deep A horizons that are rich in organic matter supply these soils with abundant nutrients. Grassland soils often have a good texture, good aggregation and other physical properties that allow them to withstand some of the negative impacts of frequent tilling.

One of the most limiting factors to crop production of these lands is the availability of water. Some grasslands have adequate water for crops such as wheat, or for grazing livestock (Figs. 5.28 and 5.29), and don't need any additional water, particularly if the soils are allowed to remain fallow some years to replenish soil water storage. On floodplains and lowlands, grasslands rarely need additional water as long as the water table is close enough to the surface for deeper roots to access it (Fig. 5.28).

Floodplain grasslands soils can have too much water and need drainage to be productive. The roots of most crops need oxygen, and a soil that is too wet can run out of oxygen (Fig. 5.30). A lack of oxygen means that roots do not have energy to take up nutrients. To optimize productivity, a soil needs to have a good balance of available moisture and still have oxygen. Thus, some agricultural soils have drainage systems installed to make sure that air can enter the soil. A drainage system can be as simple as a series of ditches that water can flow into and away from a site.

Wheat, rice, and corn (or maize) make up more than 60 % of the food we eat. Vast areas of land and soil are devoted to grain production. Globally, almost 7 million hectares of land are used to produce grains, with some of our best soils used for grain production. For example, a large area of the Central United States has excellent soils that developed under prairies. Much of this area is called "the corn belt" and produces about 40 % of the world's corn. Corn is a demanding crop requiring a lot of nutrients and water, so it is often rotated with soybeans to replenish soil nitrogen and often requires

Fig. 5.28 A rich floodplain soil with cabbages growing in western Washington State, USA. The deep A horizon is above an older deposit of alluvium. The underlying soil has mottling as it is wetter and closer to the water table. The red soil above the mottles is from a prehistoric fire

Fig. 5.29 Intensive sheep grazing on a productive volcanic soil with a moist climate in New Zealand

Fig. 5.30 Pumpkins growing on a floodplain soil that has water ponded on the surface. The fine texture of this soil and its location near a river mean that heavy rains can easily saturate the soil

Fig. 5.31 A productive field of corn

irrigation. Corn is used to feed livestock more than it is used directly as food for people, and now it is also used to produce ethanol for fuel. Having such immense areas of land covered with one crop is a concern for the long-term sustainability of these soils, ecosystem biodiversity, and water quality. However corn belt soils have been producing crops for over 150 years providing food for generations. Corn (or maize) is grown on many soils at many locations throughout the world (Fig. 5.31).

Over 700 million tons of wheat is produced worldwide every year. Wheat can grow on many soil types and tolerates a broad range in rainfall (largely due to the many different varieties of wheat). Wheat can be found growing from lower elevation lands to drier high-elevation lands (Figs. 5.32, 5.33, and 5.34). Wheat is grown on more land than any other crop, with over 200 million hectares of land used for wheat production globally.

Some rangelands are used for crops such as wheat that can grow with less moisture, but most rangelands are used for grazing livestock. Grazing animals can be kept within feedlots, fenced pastures, or open rangelands. Feedlots are fenced areas where densely packed animals are fed grains that are brought to them, typically soon before being shipped to market. The animal waste from feedlots can damage soil and pollute nearby waters if there is runoff, so careful management of feedlots is necessary to prevent environmental problems. In intensive pastoral grazing systems, animals are often rotated among fenced pastures to maximize pasture and animal productivity. Open rangelands usually have livestock rotated among different areas as well to prevent overgrazing. Drier rangelands support limited livestock as they have less vegetation to feed animals and are more susceptible to erosion if too much vegetation is lost. Desertification happens when drier rangelands become sparsely vegetated like a desert—if too much vegetation is lost through overgrazing then erosion can remove the topsoil preventing plant regrowth. Within rangelands, moister valleys support denser vegetation and more

Fig. 5.32 Wheat production in a temperate area of the Western United States. Note that the fields are plowed along the contour to reduce erosion and some fields are left fallow to increase soil moisture

Fig. 5.33 Wheat production in two dry, high-altitude environments of Peru; wheat straw can be used for livestock

intensive grazing (Figs. 5.35). Using areas near streams for livestock can cause increased erosion and allow animal wastes to enter streams so riparian areas are often protected.

People have used many types of land to graze animals for thousands of years including alpine meadows and the high-elevation plateaus of Tibet and the Andes mountains, as well as the Arctic tundra (Fig. 5.36).

Fig. 5.34 Although wheat (*left*) is the dominant crop rotated in the Australian wheat belt of Western Australia, it is rotated with lupine (*center*) and canola (*right*). Crop rotation is an important part of cropland management to minimize disease buildup and help retain soil fertility. The lupine, soy beans, or alfalfa are used to fix nitrogen and provide seed and fodder, and canola is used to produce oil

Fig. 5.35 Rangelands of Southern California, USA. *Left*: managed (background) and unmanaged (foreground) lands from a semiarid rangeland area. *Right*: moister rangelands near a valley bottom in the foothills of the Sierra Nevada Mountains

Fig. 5.36 Traditional grazing lands cover a wide range of environments. *Left*: Swiss alpine meadows where cowbells are used to keep track of animals. *Center*: nomadic herders bring their yaks and other animals to graze on the Tibetan Plateau in the summer. *Right*: sheep in a highland area of northern Spain

Chapter 6
Soils in Extreme Environments

The landscape of Victoria Valley in the McMurdo "Dry Valleys" of Antarctica is a cold desert where soils contain ice-cemented permafrost and patterned ground is evident at the surface. The Victoria Glacier is visible in the distance

© Springer International Publishing Switzerland 2016
M.R. Balks, D. Zabowski, *Celebrating Soil*, DOI 10.1007/978-3-319-32684-9_6

Introduction

"Why are we so susceptible to the charm of these landscapes when they are so empty and terrifying?"

Jean-Baptiste Charcot (A French polar scientist who visited the Antarctic in the 1900s).

Extreme environments limit life. In this chapter, we examine the landscapes and soils that form in Earth's most extreme environments and the organisms that survive there. Extreme environments are generally limited by moisture and temperature and include both hot deserts such as the Sahara and cold deserts such as the Dry Valleys of Antarctica. Geothermal activity creates extreme environments of heat and acidity or alkalinity that challenge survival for all but a few specially adapted organisms.

Hostile environments are not the place for productive agriculture, but their raw beauty is inescapable, and there are fascinating lessons to learn when you can watch soil processes stripped bare of the impacts of much of the life that marks other environments.

The soils that form in extreme environments are surprisingly diverse. They develop in a wide variety of parent materials, ranging from solid rock exposed by erosion in high mountains to acid-infused clays in geothermal environments. Over time, mineral rock materials weather to form soils, even in the most inhospitable environments.

In extreme environments, humans are only brief visitors as we cannot survive the excessive heat or cold without recourse to food and materials from other, more inhabitable environments. However, specially adapted organisms, from thermophilic (heat loving) microbes in geothermal areas to penguins in Antarctica (Fig. 6.1), are intriguing inhabitants of our extreme environments.

Fig. 6.1 Curious penguins check out a soil profile on the coast of Antarctica

Antarctic Coastal Margins

By Antarctic standards, the coastal margins are relatively warm, with mean annual temperatures close to 0 °C in the northern climes of the Antarctic Peninsula (which lies to the south of South America) and around minus 18 °C along the margins of the Ross Sea (to the south of New Zealand). The coastal margins generally have higher snowfall than inland regions of Antarctica. Occasionally, in summer, it is even warm enough to experience rainfall. The coastal margins are important as they provide pockets of snow and ice-free land (Fig. 6.2) which provides refuges where birds, such as some penguin species, can nest. Humans also prefer the ice-free areas to establish their Antarctic bases.

The coastal margins of Antarctica are fairly inhospitable environments with many areas subject to strong winds and long periods of cloudy weather in the summer time, as well as several months of darkness in winter. However, in 6 weeks or so of high summer, soil surface temperatures can rise as high as +20 °C on sunny days, and snow and glacier meltwaters provide moisture that supports some plant life. In the Ross Sea region, there are no flowering plants or conifers and thus the most evolved plants are mosses. On the relatively warmer Antarctic Peninsula, two species of flowering plants occur, along with a range of mosses and liverworts.

Many of the soils in Antarctic coastal regions are formed in a mixture of sands, gravels, and boulders (Fig. 6.3) that have been deposited by glaciers. The Antarctic coast is generally protected by sea ice (the frozen sea surface) for most of the year, but for a few weeks in late summer, the sea ice breaks up and melts, exposing much of the coast to the open sea and the soils to inputs of windblown salty sea spray.

In winter, the soils freeze as hard as concrete. In summer, the surface layers of the soil thaw. The depth to which the soil thaws is called the active layer. Near McMurdo Station and Scott Base, at a latitude of 77 °S, the maximum summer thaw depth is usually between about 10 and 30 cm. In sheltered sunny areas and in warmer areas further north such as Cape Hallett at 72 °S, the depth of thaw can be up to 1 m. Below the maximum summer thaw depth, the soil is permanently frozen, forming permafrost. In the moist coastal zones, the permafrost is strongly ice cemented making excavation difficult without recourse to jackhammers or bulldozers.

Patterned ground is common in Antarctic coastal areas where the soil has a moderate moisture content (Fig. 6.4). In the extreme cold of winter, the ground shrinks and contraction cracks occur,

Fig. 6.2 Coastal margin in the Ross Sea region of Antarctica. The Wilson Piedmont Glacier creates an inland barrier on the left of the picture. Sea ice and icebergs in the Ross Sea are visible to the right. Pockets of snow accumulate and melt intermittently during the summer, providing the soils with some moisture. Meltwater from the glacier also feeds a series of small lakes and streams in summer. The distance from the glacier to the coast is about 3 km

Fig. 6.3 Typical soil
near Marble Point in the
Ross Sea region of
Antarctica formed in
glacial till. The surface
has some salt evident
with a moderately
developed desert
pavement. The gravelly
sandy soil contains
many pebbles, boulders,
and large rocks

Fig. 6.4 Patterned ground. *Left*: view of patterned ground from the air. Each polygon is about 10 m across. *Right*: a vertical road cutting has exposed an ice wedge in a patterned ground crack that is about 2 m long. The top of the ice wedge marks the top of the permafrost. The material above the permafrost, called the active layer, is subject to seasonal thaw

much like the cracks in a drying mud puddle but on a larger scale. The patterned ground polygons are often about 10 m across and form in varying patterns depending on the soil materials and conditions. Snow is often blown across the ground surface and it tends to get caught in the open cracks. Some of the snow melts and seeps down into the cracks until it refreezes. Gradually, over many seasons, the amount of moisture in the cracks builds up, forming wedges of ice.

Fig. 6.5 Human activity in Antarctica is concentrated near the coast. *Left*: Scott's Hut at Cape Evans was built in 1911. *Right*: McMurdo Station, a US base on Ross Island, is the largest base in Antarctica

Human activity is concentrated on the coastal margins of Antarctica. Most access is by ship; thus, ice-free areas near the coast are preferred for buildings. The first huts were built by early explorers such as Scott and Shackleton (Fig. 6.5) in the early twentieth century. Since the late 1950s, many bases have been established around the Antarctic coastline, the largest of which is McMurdo Station on Ross Island (Fig. 6.5).

In some areas, the subsurface contains a large amount of ice, with just a thin covering of soil on the surface that insulates the underlying ice from melting (Fig. 6.6). There is rock material within the ice, so if the ice is exposed at the surface, the ice will sublimate (evaporate without a meltwater phase), or melt and evaporate, until enough mineral material has accumulated at the surface to form a new insulating layer. If the substrate is ice rich, then the surface may subside quite a lot before enough material has accumulated on the surface to insulate the underlying ice from further melting or sublimation.

At some Antarctic bases, material from the summer-thawed active layer has been repeatedly removed to establish flat building sites or to provide fill material for adjacent areas. Where the active layer has been removed, a new active layer becomes established as the top of the underlying ice, which was formerly within the permafrost, is melted and evaporated. If the upper layer of the permafrost is ice rich, there may be quite a lot of melting, removal of water, and surface slumping before enough mineral material accumulates at the surface to insulate the underlying ice from further melting and a new, stable surface forms.

Some of the permafrost ice has a high salt content, mainly derived from nearby seawater. In coastal areas, in the Ross Sea region, when the underlying permafrost melts, after the surface active layer has been removed, then the melted water evaporates from the ground surface leaving a buildup of salt behind (Fig. 6.7).

Some of the more unusual soils on Earth are formed in penguin colonies such as the Adelie penguin colony at Cape Hallett (Fig. 6.8). Adelie penguins build nests from small stones in order to make a platform that keeps eggs and chicks above the influence of meltwaters. The preferred stones are about 2–6 cm in diameter, big enough to be useful in forming a nest and small enough for the penguins to pick up in their beaks. To Adelie penguins, stones are more valuable than gold and much time is spent fighting over them, guarding them from marauding neighbors and searching for new stones to enhance their nest. The demand is ongoing as the nests also accumulate liberal doses of guano and occasional remains of dead birds. Thus gradually a soil is developed in the buildup of stones and guano. Carbon dates of penguin remains show that the penguins have been nesting at Cape Hallett for at least 1000 years, and in that time, about 1 m of "soil" has accumulated.

The soils formed in penguin guano are unique—think of the floor of a hen house that has been there for 1000 years (Fig. 6.9). The soil is very sticky and very smelly! The soil at Cape Hallett (after

Fig. 6.6 Ice-rich permafrost exposed by stream erosion near McMurdo station in the early 1990s. The insulating coating of mineral material at the soil surface (the active layer) is clearly visible

Fig. 6.7 *Left*: an area adjacent to McMurdo Station on Ross Island that has been "scraped." The surface summer-thawed (active) layer was removed to use as fill material. The newly exposed underlying ice has melted and evaporated leading to surface slumping and salt accumulation on the new soil surface. The white material on the ground surface in this photo is salt, not snow

Fig. 6.8 A small part of the penguin colony on the coast in the Ross Sea region of Antarctica, where thousands of penguins nest in summer

Fig. 6.9 Soil formed from penguin nesting stones and guano is exceptionally sticky and stinky. *Left*: the soil profile. *Center*: the penguin colony. *Right*: a penguin on its nest of stones

removing stones) is about 12 % phosphorus—high enough to be useful as fertilizer. The guano-derived soil also has high levels of elements such as arsenic and cadmium which are likely to be responsible for the relatively low numbers and diversity of microbes in these soils.

The Dry Valleys of Antarctica

The Antarctic Dry Valleys have stunning landscapes, a fascinating array of soils, and some of the most extreme environments on Earth (Fig. 6.10). Some space scientists study the Dry Valleys as they are considered the environment on Earth that is most similar to Mars. The Dry Valleys are "dry" (i.e., largely free of snow and ice), because here evaporation exceeds precipitation. Precipitation (as snow) is exceptionally low as the main source of atmospheric water, the open sea, is some distance away. Evaporation is enhanced by strong winds that flow down off the Polar Plateau through the Dry Valleys.

The initial impression of the Dry Valleys is one of rocks, but if you dig a little deeper, a surprising variety of soils occur. The soils within the Dry Valleys vary from very young soils on active sand dunes, and the floodplains of the Onyx River to soils that some consider to be millions of years old formed on stable surfaces at higher altitudes that have largely escaped the effects of ice advances over the last glaciation.

Fig. 6.10 The North Fork of the Wright Valley viewed from The Labyrinth. The Wright Valley, at over 50 km long, is one of the longer Dry Valleys. Lake Vanda is evident in the middle distance, and small alpine glaciers that flow partway into the Wright Valley from the Asgard Range are visible in the far distance

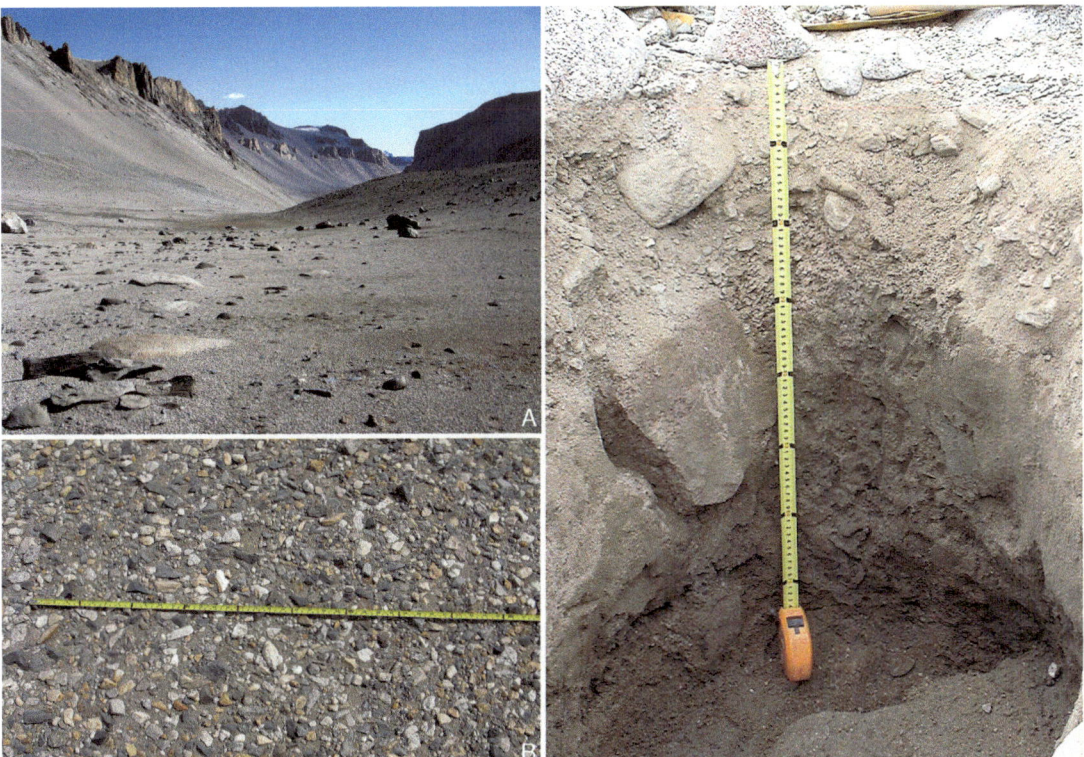

Fig. 6.11 The Dry Valleys. *Top left*: view of the valley floor in the Wright Valley where glacially deposited rocks are scattered across the surface. *Bottom left*: looking down on the surface desert pavement. A variety of different rock types are evident, including gneiss, granite, dolerite, and sandstone. *Right*: typical unsorted rocky, gravelly, sandy soil formed in glacial till. The temperature in the base of the hole is about −14 °C, and the dry permafrost extends to within about 30 cm of the soil surface (The *dark marks* on the tape are at 10 cm intervals)

    Soils on the floors of the Dry Valleys are generally formed in glacial tills and are dominated by rocks, gravel, and sands (Fig. 6.11). In areas where there is no adjacent water source (such as a zone of meltwater runoff or a lake margin), the soils are so dry that there is not enough moisture to form ice cement, even where the soil temperature is well below 0 °C. Thus, it is relatively easy to dig holes to explore the soils and underlying permafrost. A desert pavement forms on the soil surface with an armoring of gravel and stones protecting underlying fine material from blowing away.

    At higher altitudes, many of the soils are considered to be formed on surfaces that have not been greatly disturbed for millions of years (Fig. 6.12). While weathering processes are extremely slow in the cold dry environment, the huge time span gives rise to some marked differences in comparison with the soils at lower elevations that are formed in more recently deposited materials. Oxidation of iron-rich minerals gives soils redder colors, and salts, derived predominantly from rock weathering, have often accumulated to form distinctly salty layers within the soil.

    The valley walls comprise a series of scree slopes and fans where material moves downhill, accumulating toward the base of the slope. While it is easy to think that the Dry Valley landscapes are effectively "frozen in time," in fact, some of the colluvial fans are quite active (Fig. 6.13). Meltwater flows down many of the fans every summer, sometimes finding its way beneath the surface, occasionally as small ephemeral streams that flow for perhaps just 2 or 3 weeks each year. Where a larger-than-usual melting event occurs, water erodes material and moves it from high on the valley walls down to form fans on the lower slopes.

Fig. 6.12 A high-altitude valley. The patterned ground is marked by snow caught in the patterned ground cracks. The soil here is strongly weathered compared to younger, lower altitude, soils. The pale colors on the face of the excavated pit are salts accumulated due to evaporation of water from the soil in the hour or two since the pit was dug

Fig. 6.13 *Left*: A colluvial fan near Lake Vanda that was active in about 2008. New gully erosion has occurred in the upper part of the slope, and fine material has been carried down and deposited over the ground surface on the foot slope. *Center*: typical soil profile on a colluvial fan in the Wright Valley where horizontal layers of gravel and sand reflect the periodic erosional and depositional events that have gradually built up the fan material. The shallow soil profile is the result of ice-cemented permafrost making it impractical to dig below about 30 cm depth. *Right*: looking down a frozen ephemeral stream that is fed by glacier meltwater. This stream flows (intermittently) into the Onyx River on the valley floor. The darker gray stream margins are the result of soil moisture that seeps out from the streambed

Most striking of all are the desert pavements on old surfaces (Fig. 6.14). The strongly weathered desert pavement rock surfaces are highly polished and pitted as a result of long periods of exposure to wind and its attendant sandblasting. Temperature extremes, salt crystallization, and freezing and thawing of water contribute to the weathering and development of desert pavement surfaces.

Water plays an important role within the desert environment of the Dry Valleys. The U-shaped valley cross sections are a reminder that it was the erosive forces of huge, long-gone glaciers that originally carved these valleys. Where water moves, it also carries salt, dissolved from soils and concentrated and recrystallized when the water evaporates. An exceptional example is Don Juan Pond (Fig. 6.15).

Don Juan Pond contains the saltiest natural water on Earth and is so salty that the water does not freeze solid, even at temperatures as low as −40 °C in the Antarctic winter. Recently researchers have

Fig. 6.14 Weathered rocks in a high-altitude desert-pavement surface. *Left*: a stone surface pavement. *Right*: the small stones and sand that lie in the pits are likely to be moved around by the actions of wind and frost to gradually enlarge the hollows and slowly wear the rock down over thousands of years. The polished nature of the surface is an effect of wind and weathering known as "desert varnish"

Fig. 6.15 Don Juan Pond in the South Fork of the Wright Valley. The white "pond" is not formed of snow or ice but salt crystals. Water seeps to this low point in the landscape then evaporates, leaving the salt behind

Fig. 6.16 A small lake in Taylor Valley viewed from a helicopter. The halo effect of moistened soil on the lake margin, with salt accumulating on the outer edge as water evaporates, is clearly visible. The *brown* zone nearer the water's edge is likely to be cyanobacteria growing in the periodically wetted lake margin

suggested that some of the water that helps move salt into the pond, and keep it liquid, is absorbed by the salts from the humidity in the atmosphere. They think similar processes may occur on Mars.

While Don Juan is an extreme example, many other lakes in the Dry Valleys have significant salt contents. Lake margins have a range of conditions of varying moisture and salinity that influence the soils formed (Fig. 6.16).

Furthest South: Soils on the Margins of the Polar Plateau

Some of the most southerly soils on the planet are about 700 km from the South Pole, at about 85° south, on the margins of the Polar Plateau. The Polar Plateau is an expanse of ice over 1000 km across, with an average altitude of about 3000 m above sea level. Ice has accumulated slowly on the Polar Plateau and gradually overflows around the margins of the Transantarctic Mountains, forming some of the largest valley glaciers on the planet. The Beardmore is one such glacier. Up to 40 km wide, the Beardmore flows 201 km from the Polar Plateau to the edge of the Ross Ice Shelf. The Beardmore Glacier was famously first traveled by Ernest Shackleton in 1908 and then in 1911–1912 by Robert Scott on his ill-fated expedition to the South Pole. As the Beardmore Glacier has flowed down from the Polar Plateau, along the edge of the Dominion Range, it has scraped rock material off the edges of the mountains and deposited it to form a sequence of lateral moraines along the margins of the Range (Fig. 6.17).

Fig. 6.17 Lateral moraines make linear features across a 4 km wide swath in the landscape between the foot of the Dominion Range and the edge of the Beardmore Glacier. The Beardmore Glacier merges into the Polar Plateau on the horizon toward the right-hand side of the picture

At ground level, the orderly landscape pattern we see from the helicopter is harder to discern, with the ground surface appearing as a lumpy conglomeration of rocks that form "patterned ground" with an incipient desert pavement. On the areas with lateral glacial moraine sequences, a distinct pattern is evident in the soils. Near the glacier margin, there is only a thin veneer of mineral material overlying massive glacial ice (Fig. 6.18). The soils become older and deeper as one moves away from the glacier edge (Figs. 6.19 and 6.20), reflecting the gradual removal mainly by sublimation of underlying ice, leaving behind the rock material, which was caught up in the ice. In this environment, at about 2000 m altitude, even at the height of summer, the soil surface does not reach 0 °C (summer air temperatures are around a chilly −15 °C) and there is minimal precipitation. Although the permafrost reaches to the soil surface, the soils are so dry that there is no ice cement.

The pattern of lateral moraines, with shallow soils near the glacier edges and deeper soils formed inland, is repeated at a number of sites in the region including near Mt. Achernar (Fig. 6.19) and the Ong Valley (Fig. 6.20).

Perhaps the most extreme soil environment in this region is on the Otway Massif, a small outcrop of rock at an altitude of over 3500 m against the edge of the Polar Plateau at the head of the Beardmore Glacier. The exposed ground surface is considerably older (probably millions of years old) than the glacial moraine sequences discussed above and is underlain by some of the oldest soils anywhere on Earth. The soils (Fig. 6.21) are strongly weathered by Antarctic standards. Although soil-forming processes are incredibly slow, there has been immense time for soil development. Sodium nitrate salts have accumulated, gradually being deposited from aerosols in the atmosphere.

Fig. 6.18 Soils between the Dominion Range and the Beardmore Glacier. *Left*: the broad color strips across the lateral moraine sequence reflect the domination of various rock types eroded and carried in by the glacier from different rock outcrops. *Right*: near the glacier edge, there is only about 10 cm of soil over glacier ice

Fig. 6.19 Mt. Achernar area, midway across the moraine sequence. *Left*: As you move inland, away from the glacier edge, the desert pavement becomes more strongly developed. The line of pale material along the side of the cliff face marks the height to which the last glacial advance reached. *Right*: a soil profile from several kilometers inland comprises 50 cm of loose gravelly sand over glacial ice with a moderately developed desert pavement near the surface

Fig. 6.20 Ong Valley. *Left*: view from a helicopter looking toward the head of the valley. The patterned ground mounds are generally about 20 m across. *Right*: deep soil profile on an ancient terminal moraine near the head of the valley has a *brown color* indicating some weathering of iron minerals and dry permafrost with no ice cement present to at least 80 cm depth

Fig. 6.21 Landscape and soil on the Otway Massif, one of the most extreme, ancient, and southerly soils on Earth, at 3500 m altitude and 500 km from the South Pole. The soil contains sodium nitrate salt that has been deposited from the atmosphere

Mid-Latitude Deserts

Deserts have limited water and so vegetation is generally sparse. Thus the exposed rock materials are subject to the impacts of extreme temperatures, wind, and rare rainfall events, which shape the landscape and the soils. The lack of water limits chemical weathering, and hence physical weathering processes tend to dominate. Physical weathering breaks rock materials into smaller pieces, without chemical change, thus resulting in the formation of sand-sized material and our general perception that deserts are mainly sand (Fig. 6.22).

Fig. 6.22 Sand dunes such as these in Death Valley, USA, are common in desert environments where there is no vegetation or stone pavement to protect soil from wind erosion and redeposition

In deserts several physical processes wear away at the rocks to form soil materials. Abrasion from wind-driven sandblasting wears surfaces down, polishing hard rocks that are resistant to weathering and gradually chipping small pieces off less resistant rocks. Extreme temperature fluctuations can cause the outer part of a rock to expand faster than the internal part, thus creating forces in the rock that lead to cracks or fissure forming. Salt crystals can form at rock surfaces and force mineral grains apart.

Over long periods of time, on exposed desert surfaces, erosion gradually moves materials, via wind, and occasional rainstorms. Sand dunes may form and gradually move across the landscape. Where some parts of the landscape are resistant to erosion, interesting landscapes develop. Mesas and buttes (flat-topped erosion remnants that stand above the surrounding landscape) form where a resistant layer protects part of the land surface from erosion while material is gradually removed from the adjoining landscape (Fig. 6.23).

Arid regions, such as those of Nevada, Arizona, and Utah in the USA, are defined by lack of water, summer heat, minimal vegetation, and shallow sandy soils with little organic matter. However water has a huge influence on the landscapes that form. Water erosion occurs when occasionally heavy downpours fall on the weakly vegetated and thus largely unprotected, land surface. Large rivers, such as the Colorado, flow through the area with water fed from distant mountains and have gradually cut huge canyons across the landscape (Fig. 6.24). Arches are spectacular features that occur where some parts of the geologic deposit weather and erode more readily than others (Fig. 6.24).

Fig. 6.23 The desert landscape in Monument Valley, USA, is shaped by erosion that has removed material leaving only small remnants of the former land surface which form features known as buttes. Over time these too will succumb to erosion until the surface becomes uniformly flat. The "salt bush"-type vegetation here is tolerant of dry and salty soils but indicates that enough rainfall occurs to sustain some plant life

Fig. 6.24 Desert landscapes where weathered soil materials are removed by water. *Left*: the junction of the Green and Colorado Rivers where and deep canyons have been carved by the two rivers. *Right*: Delicate Arch, Arches National Park, USA

In desert regions, where the potential for evaporation greatly exceeds the rainfall, salts accumulate (Fig. 6.25). In some areas, after heavy rain, the water runs into low-lying areas carrying salt with it then evaporates leaving the salt behind. Huge salt lakes or playas can form. Salt renders the soil uninhabitable for all but the most extremely salt-tolerant plants. Because desert soils have little protection from vegetation occasional rainfall, events can lead to runoff and rapid erosion (Fig. 6.26).

Fig. 6.25 A broken salt pan surface in Death Valley known as the "Devil's Golf Course." Finding golf balls in this environment would be a challenge!

Fig. 6.26 An eroded gully where water flows after an occasional rainfall event on the margin of the Gobi Desert in China

Extreme Heat: Geothermal Environments

Geothermal areas occur where volcanic activity provides a source of steam and superheated water near the Earth's surface. In this hot, wet, and often extremely acid or alkaline environment, chemical weathering reaches unusual peaks and provides a rich palette of mineral assemblages that form unique and beautiful, if often inhospitable, soils. In geothermal systems, groundwater comes in contact with subsurface heat. Deep within the ground, the water is under pressure so it can reach temperatures much higher than 100 °C before turning to steam. Where the superheated water finds its way to the surface features such as geysers and if the opening is larger, pools of hot water form (Fig. 6.27). The water is usually alkaline, and chloride often dominates the chemical makeup of the waters, so they are referred to as alkali-chloride.

When superheated water seeps through the ground, it dissolves rock minerals and carries the elements in solution. As the water moves toward the ground surface, the pressure and temperature gradually drop, and different elements precipitate out of solution in different zones. Copper, lead, and zinc oxides tend to precipitate out relatively deep beneath the ground surface, forming concentrated deposits as the waters flow over long time periods. Gold, silver, and uranium precipitate out at temperatures and pressures that occur just below the ground surface; thus, they become concentrated near the ground surface. As the water rises to the surface elements such as antimony are precipitated and as the water cools to surface temperature, flowing away from the geyser or pool, silica is deposited, often forming extensive silica sinter (welded crust) terraces. Much carbon dioxide, along with volatile

Fig. 6.27 Alkali-chloride geothermal features at Yellowstone National Park in the United States. *Left*: Old Faithful Geyser. *Top right*: a pool where alkaline water is seeping to the surface and overflowing. The varying colors around the edges reflect different minerals precipitating out as the water temperature gradually drops. *Bottom right*: extensive silica sinter is deposited where geothermal waters overflow

Fig. 6.28 Acid-sulfate geothermal features. *Left*: boiling mud, Rotorua, New Zealand. *Right*: soil material that has been chemically weathered due to geothermal steam, Hawaii. The *yellow* colors are deposits of sulfur, *dark red* is hematite (an iron oxide mineral that can form at high temperatures), and the *white* materials are gypsum (calcium sulfate) and silica

Fig. 6.29 Mt. Erebus in Antarctica. *Left*: the mountain viewed from Scott Base on the Ross Ice Shelf is an active volcano with a constant plume of steam rising from it. *Right*: the crater at the top of Mt. Erebus where the steam heat prevents snow and ice from accumulating (Photo: Tanya O'Neill)

elements such as mercury, and of course water vapor, is released to the atmosphere as geothermal waters rise to the surface. Many mineral deposits that are mined around the world were concentrated by ancient, now cold, geothermal systems.

Where the pressure drops within the ground, often around the margins of alkali-chloride zones, the hot water flashes to steam and the steam then seeps up through the ground where it interacts with the soil to form acid environments. Elements such as sulfur precipitate out of solution. Boiling mud occurs and occasional steam explosions form small steamy craters (Fig. 6.28). In acid-sulfate environments, the sulfur reacts with water forming sulfuric acid, and the pH can drop as low as about three. Only the most acid-tolerant plants and microbes survive such conditions.

The ultimate extreme geothermal environment is at about 3700 m above sea level on the top of Mt. Erebus in Antarctica (Fig. 6.29). Here the atmosphere is so thin that helicopters cannot safely fly to the top, the air temperature, even in mid-summer, is around −25 °C, and yet geothermally heated soils reach temperatures of up to +65 °C. Scientists have found microbes surviving in this most isolated and extreme environment. The soils on Erebus are heated by steam from fumaroles. There are very strong temperature and pH gradients. Soils near the fumarole margins have a near-neural pH, while a few meters away, influenced by the sulfur in the steam, soils have pHs as low as 4.

Celebrating the Productivity of Soils in Extreme Environments

The productivity of soils in the most extreme environments is generally limited by water availability or temperature and sometimes also by extremes of acidity or alkalinity. Despite these challenges, life persists and a diversity of specialized organisms survive.

Where there is liquid water, there is life. Nowhere is this more evident than in the Antarctic. Over the summer, in the coastal Antarctic, meltwater seeps from snow patches and glaciers. Thus, in places, you find small ephemeral streams that thaw and flow for a few hours during the day. *Cyanobacteria* are quick to become established in ephemeral streams and can create extensive mats where flat surfaces lead to wide shallow water flow. In areas that are wet with shallow water for longer periods, mosses may become established, growing for just a few weeks per year (Fig. 6.30).

Over the majority of the ice-free regions of Antarctica, liquid water is minimal, so the landscape is largely barren with no visible evidence of plant life. However, even in the most hostile Antarctic environments, microbes survive in the soil. New DNA measurement techniques allow scientists to learn more about the microbes in Antarctic soils. While not as diverse or abundant as soil-microbial populations in warmer regions, there are many thousands of microbes present in even the driest, most extreme, soil environments. There is a short food chain which includes bacteria and tiny nematodes (microscopic worms, Fig. 6.31). In some areas, tiny red mites and small springtails can just be distinguished with the naked eye. Microscopic photosynthesizing cyanobacteria find niches within rock surfaces where there is a little moisture and light (Fig. 6.31).

When rare rainfall occurs in mid-latitude deserts, a wide range of plants and other organisms are quick to take advantage of the moisture. Localized showers of rain initiate a rapid plant growth spurt. Tiny plants spring up, flower, set seed, and then die away to await the next rainfall event, which may be years away (Fig. 6.32). Caterpillars also hatch to take advantage of the brief feast.

Hardy plants, such as cacti, are adapted to long periods of drought and hot desert environments (Fig. 6.33). Extreme environments are characterized by extreme temperatures and are also often limited by water availability. Water may be frozen solid, too salty, or simply absent. Thus observing the plants, animals, and microbes that survive in the most extreme environments on Earth provides us with a sense of awe at the tenacity of life and the unlikely combination of simplicity and complexity in the soils that develop to support life in such arduous conditions.

Fig. 6.30 Moss and lichens are both found in moist environments. *Top*: moss beds occur where there is extensive melt-water runoff at Cape Hallett in Antarctica. The penguins contribute nutrients to the environment. *Bottom left*: moss grows strongly for a few weeks when moisture conditions are favorable then lies dormant for many months. *Bottom right*: lichens occur in tiny isolated patches throughout the ice-free areas of Antarctica. However in the warmer, wetter environment of Cape Hallett lichen growth is positively "lush" by Ross Sea Region standards

Fig. 6.31 Even in the cold desert soils of the Antarctic Dry Valleys some microbes survive. *Left*: a microscopic nematode, known as Scottnema lindsayae, is relatively common. (Photo: Diana Wall). *Right*: The green tinge on the underside of an overturned stone is evidence of microbes surviving under this rock where a little light seeps in and some moisture is present

Fig. 6.32 A wide variety of miniature flowering plants quickly grow, flower, and then die following rainfall in Death Valley, USA. *Right*: tiny plants are being eaten by a hungry caterpillar that has also grown after the rain. A car key provides a scale

Fig. 6.33 Cacti, adapted to harsh desert conditions, come in a wide range of forms. *Left*: a flowering cacti in the Southern California desert. *Right*: a Joshua tree in Nevada, USA

Chapter 7
When the Earth Moves: Earthquakes, Landslides, and Erosion

A person negotiates a road that has been repeatedly damaged by landslide activity

© Springer International Publishing Switzerland 2016
M.R. Balks, D. Zabowski, *Celebrating Soil*, DOI 10.1007/978-3-319-32684-9_7

Introduction

> "Time brings change. Even the mountains, that seem to stand for permanence when measured against our brief stay on the planet, are in a state of flux, rising and being worn away."
>
> Tristan Gooley 2012.

We sometimes describe the land as "terra firma"—the Earth seems solid and safe—but in some parts of the world, we can find ourselves standing, literally, on shaky ground! The major tectonic processes on planet Earth are largely responsible for keeping land above sea level and, thus, make terrestrial life possible. Over long time periods, whole continents are created and destroyed. Mountain chains are uplifted, eroded, and washed back to the sea. Faults, such as the San Andreas Fault on the west coast of North America and the Alpine Fault in New Zealand, occur where huge tectonic plates move past one another and, over time, create offsets of hundreds of kilometers.

On a more human timescale, we deal with the day-to-day effects of earthquakes and landslides. Major Earth movement events, whether as a result of landslides or earthquakes, impact human infrastructure and affairs. Earthquakes are particularly terrifying with faults causing direct land surface disruption. Earthquakes also trigger landslides, and the ground shaking can cause sandy soils to liquefy with sand "boiling" up at the ground surface. Humans have no control over earthquakes, but we can identify areas at risk. As populations and infrastructure grow, the potential impacts of earthquakes become more devastating. There is much we can do to understand the likely effects of earthquakes and to build infrastructure that can withstand strong shaking and minimize human losses.

Steep lands, especially when combined with high rainfall, are prone to landslides and erosion (Fig. 7.1). Globally, landslides kill hundreds of people every year. As the human population has grown, more people are living on steeper slopes and are thus more prone to suffering the adverse effects of moving Earth.

Fig. 7.1 Saturated soil has become part of a mudflow. Stream erosion is removing material from the foot of the slope and thus rendering the hillside above unstable (area about 2 × 1 m)

Uplifted Land: New Opportunities

Collisions between tectonic plates often result in land being pushed upward. One of the most rapidly uplifting areas is the Himalaya Mountains where the Indian tectonic plate is colliding with the Eurasia tectonic plate. The Himalayan Mountains are being uplifted at an average rate of about 1 cm per year resulting in the highest mountain peaks on Earth. The process goes on with many small increments such as that of the Kunlun Shan earthquake on the margin of the Tibetan Plateau in 2001 (Fig. 7.2). Erosion is also an important process in mountains such as the Himalaya, gradually removing material. It is a balancing act between the rate of uplift and the rate of erosion that determines the height of the mountains.

New Zealand provides us with some interesting examples of uplift. The Southern Alps are uplifting at an average of 1–5 mm per year and much (but not all) of the rest of New Zealand is also being uplifted, though at slower rates. There have been two major uplift events in New Zealand, in the last 200 years, one near the capital city, Wellington, and the other 300 km north near Napier.

In 1855, a magnitude 8.2 earthquake caused widespread uplift centered on a fault rupture about 20 km east of Wellington, New Zealand. There was up to 6 m of vertical uplift and 18 m of horizontal offset. Wellington, on the edge of the harbor, was uplifted so that the former road around the edge of the waterfront (known as Lambton Quay) is now situated several blocks from the waterfront. People were quick to take advantage of the newly exposed land, and much Wellington infrastructure is now built on the uplifted former seafloor. Many landslides and tsunamis were associated with the 1855 earthquake. In spite of the extremely strong shaking, fewer than ten people died.

Most of the wooden-framed houses survived the earthquake, though chimneys were shaken down. There were however some advantages. The uplift created space between the coastal cliffs and the sea

Fig. 7.2 Soil ripped apart on the fault trace of the 2001, magnitude 8.1, Kunlun Shan earthquake on the margin of the Tibetan Plateau in China. The Kunlun fault was offset horizontally by several meters in 2001 in a rift that extends for over 350 km. There was a small (few centimeters) vertical uplift on the north side of the fault (the downhill side in this picture taken 5 years after the earthquake)

Fig. 7.3 The main road
and rail corridor into
Wellington, New
Zealand, was formed on
land uplifted from the
sea in an earthquake in
1855. Prior to the
earthquake, the sea came
to the foot of cliffs at the
edge of the vegetated
area in the photograph

along which the main road and rail corridors into Wellington were subsequently constructed (Fig. 7.3), and it also facilitated drainage of swampy areas in the nearby Hutt Valley, allowing farm, and later urban, development.

Another major uplift event occurred in Napier on the east coast of the North Island of New Zealand in 1931. The earthquake caused over 200 deaths and the town of Napier was demolished, first by the shaking of the earthquake and then by fires that raged for hours afterward. As a result of the earthquake an area about 90 km long and 15 km wide domed upward. The land and seafloor were raised by as much as 2.7 m. An area that was formerly an island surrounded by tidal mudflats was uplifted so the island is now a peninsula (Fig. 7.4). An area of about 2000 ha that had formerly been coastal wetlands and a large tidal lagoon were raised creating new land.

There are some interesting challenges associated with developing soils that were once part of the seafloor. The initially uplifted sediments were high in phosphorus (from bird droppings) and calcium carbonate (from shellfish) as well as sodium chloride from sea salt. Drainage was installed and rain leached the sodium chloride salt from the sediments. Initially only salt-tolerant plants grew. Gradually over time, as the salt was washed out of the soil, other plants became established.

Six years after the 1931 Napier earthquake, the first pasture grasses were sown, and by 1939, several thousands of sheep and cattle were being run on the uplifted land. There were ongoing issues with stock health which were eventually identified as a copper deficiency caused by low copper content in the soils as well as a high molybdenum content that blocked copper uptake by plants and animals. Addition of copper sulfate to the soil overcame the problem. Organic matter has gradually built up in the soils, and as sodium has been leached, clays have become less dispersive giving improved soil structure. By the twenty-first century, just 70 years after the materials were hoisted above sea level, the soil has become sufficiently well developed to support productive farmland and an airport (Fig. 7.5).

Fig. 7.4 Maps of the Napier area before (*left* in 1865) and after (*right* in 1965) the 1931 earthquake. Much of the land that now makes up Napier city, along with an area of farmland and the local airport, was tidal lagoon until the uplift of the 1931 earthquake (Maps prepared by NZ Department of Survey and Land information)

Fig. 7.5 Changes in the landscape as a result of the 1931 Napier earthquake. *Top left*: part of the tidal lagoon taken in the 1920s (the photo here was taken of an illustration in signage at the site). *Top right*: view of the same area in 2013. *Bottom left*: farmland successfully established on what was the floor of the tidal lagoon. Note the drainage installed recently, the remaining low-lying wetland area in the middle distance, and airport buildings further away. *Bottom right*: the soil formed in sand and shell materials, which, prior to 1931, was the floor of the lagoon. Pasture has been successfully established

Fig. 7.6 Grapes growing on soils formed on seabed that was lifted above sea level in the Napier earthquake in 1931, North Island, New Zealand. The winery's name, "Crab Farm Winery," is a nod to the former inhabitants of the site

A successful vineyard, known as Crab Farm Winery, has also been established on the somewhat unusual soils of the former tidal lagoon (Fig. 7.6).

On Shaky Ground: Soil Behavior in Earthquakes

The severe shaking that occurs during earthquakes can cause an effect known as "liquefaction" where the soil material behaves much like a liquid or a wobbly jelly. Soils formed on fine sand or silt sediments with high water tables are the most vulnerable to liquefaction. Liquefaction is associated with severe ground shaking causing buildings to collapse (Fig. 7.7) and sand and water to erupt at the surface, and soil materials may compact, and thus land-surfaces subside.

Earthquake damage to infrastructure such as buildings is often worse in areas prone to liquefaction than on neighboring soils formed on harder substrates such as bedrock. Liquefaction was a major factor in the damage to San Francisco's Marina District in the 1989 earthquake and to the port of Kobe in Japan in the earthquake that struck there in 1995. The earthquakes in Christchurch, New Zealand, in 2010 and 2011 resulted in extensive liquefaction which caused damage to roads, houses, and other infrastructure such as water, gas, and sewer pipes and electricity cables. Large volumes of sand "erupted" onto the ground surface, both in the initial earthquake and in aftershocks. A huge cleanup was undertaken to remove the sand from roads, school and sports grounds, and people's backyards (Fig. 7.8).

Ground shaking also commonly causes unsupported soil materials, such as those adjacent to a stream bank, to spread laterally or collapse out toward the stream. Such instability is often highly damaging to infrastructure such as bridges (Fig. 7.9).

The shaking of earthquakes often triggers landslides (Fig. 7.10). In large earthquakes, landslides are often one of the major causes of human deaths. Earthquakes may also dislodge large rocks from steep hillsides which can roll down to cause serious damage (Fig. 7.11).

Fig. 7.7   Damage to buildings in Christchurch, New Zealand, caused by an earthquake in 2011

Fig. 7.8 Sand erupted at the ground surface as a result of liquefaction during the Christchurch earthquake of February 2011. The *thin gray line* running up through the profile is the conduit by which the sand reached the surface. The former dark topsoil is now buried by about 15 cm of sand. The water table visible in the base of the hole illustrates the saturated soil conditions that contribute to mobilization of the sand when it is shaken (Photo: Peter Almond)

Fig. 7.9 The twisted remains of a footbridge after the 2011 Christchurch earthquake. The unsupported riverbanks collapsed inward as a result of the shaking damaging many bridges and water and sewer pipes

Fig. 7.10 A large Earth slump in soft sedimentary rocks in New Zealand was set off by a magnitude 6.2 earthquake in 2014. The stream at the foot of the slope was dammed and a small temporary lake formed

Fig. 7.11 A cliff face in Christchurch, New Zealand, from which large rocks were dislodged by earthquakes in 2011, leaving several houses teetering on the edge. Containers at the bottom protect the road from falling rocks dislodged by aftershocks

Landslides and Mudflows: Hillsides on the Move

Large-scale Earth movement on hillsides is readily explained: it occurs on steep slopes, soft, or shattered rock or Earth materials and is often precipitated by high rainfall or snowmelt. The water adds to the weight of material on the slope. Conversely, high water tables also lead to increased water pressure in the pores and fine cracks in the soil, which makes the overlying material more buoyant. Water also acts as a lubricant, reducing friction between soil layers. All of these factors together lead to a situation where the gravitational forces pulling the material down the slope become greater than the frictional forces holding it up, and thus the material collapses and slides downhill. In more intensively inhabited steep areas, landslides have buried whole towns.

The east coast of the North Island of New Zealand is especially prone to landslides. The region has soft mudstone sedimentary rocks. Rapid tectonic uplift results in river downcutting and steep slopes. The area is also prone to periodic extreme rainfall events when subtropical cyclones occasionally venture south bringing heavy rainfall that precipitates major landslide events (Fig. 7.12).

Once a landslide gives way, the material can move rapidly downslope, and the mixture of soil and water often becomes a dense mudflow that can carry soil and rocks, including large boulders. Another downside to major erosion events is that the eroded material is eventually deposited further downstream on river floodplains with much material also carried out to sea. In Cyclone Bola, much eroded material was deposited on the floodplain in the lower reaches of the catchment causing extensive damage to houses and infrastructure including roads and vineyards (Fig. 7.13).

Fig. 7.12 Landslide damage to mudstone hills on the east coast of the North Island of New Zealand following a severe subtropical cyclone (Cyclone Bola) in 1988. *Left*: under pasture. *Right*: under young plantation pine forestry

Fig. 7.13 A vineyard buried by mud carried down from the hills in Cyclone Bola in New Zealand in 1988

Hill slope mass movement in regions such as the New Zealand mudstone hill country is an ongoing natural geological event (Fig. 7.14). However, extensive landslide damage has been exacerbated by conversion of the native forest vegetation to shallow-rooted grasses and sheep or cattle grazing.

Landslides often have a distinct, sloping layer upon which the soil slides—referred to as a shear plane. Often the shear plane is the top of a layer of clay that is somewhat impermeable to water; thus in wet conditions, the soil above the shear plane becomes saturated, and the clay material becomes slippery, acting like a playground slide for the material above (Fig. 7.15).

Fig. 7.14 The Tarndale Slip in the North Island of New Zealand is one of the largest in the Southern Hemisphere, covering about 40 ha. The displaced sediment is building up the bed of the adjacent river, creating a wide braided bed downstream. An extensive reforestation program has stabilized some of the surrounding hillsides but not yet gained a hold on the major landslide

Fig. 7.15 Two examples of exposed shear planes from which the overlying material has slid. Mobilized, saturated soil flowed down to the streams far below

Landslides are often triggered by the removal of material at the base of a slope. Thus landslides are common along the margins of rivers that are actively downcutting, removing eroded material and forming steep gorge walls. The coastline is another area where natural erosion, from waves impacting at the base of slopes, often initiates landslides (Fig. 7.16). Wave action, especially during storm events, can cause intense and rapid erosion. People love to view water and thus there is pressure to permit construction of houses and other infrastructure along coastlines where they are often vulnerable to the effects of erosion.

Many landslides occur along coastlines such as that of British Columbia, Canada, where there is high rainfall, steep slopes, and shallow soils on bedrock. The soils are not well anchored to the bedrock, and when heavy rains saturate the soils, landslides carry both soil and trees into the sea (Fig. 7.16). When both vegetation and soils are lost from the slope, new soil must form. Where hard bare rock is exposed, soil and ecosystem development can take a long time (many decades).

Fig. 7.16 Landslides along the coastline. *Left*: a landslide on an inlet along the south coast of British Columbia. *Right*: an active landslide on a high-energy coast with soft, easily eroded mudstone rocks in New Zealand provides serious challenges to keep the coastal road open

A combination of high-energy coasts, where the sea is constantly battering the shore, and soft readily erodible materials, leads to rapid erosion, forming steep coastal cliffs and a naturally retreating coastline in some parts of New Zealand (Fig. 7.16). Here vegetation can become established relatively quickly on the soft rock surface in the favorable climate. However the ongoing erosion and retreat of the cliff face precludes much opportunity for vegetation to gain a hold.

In March 2014, a massive landslide occurred at Oso, Washington State, USA, which covered over 2.5 km², destroyed many homes, and killed 43 people. Heavy rain combined with unstable glacial deposits created a mudflow that slid downhill, across the North Fork Stillaguamish River valley and part way up the opposite side of the valley.

The rainwater infiltrated a coarse-textured glacial outwash deposit and then encountered finer-textured glacial lake sediments below. The silt and clay in the glacial lake material slowed downward water movement, and the water increased the mass of the glacial deposits. Because the lake sediments had been lubricated by the water, the weakened material gave way, causing mud, soil, and forest debris to rapidly move downhill across the river valley leaving destruction within the landslide's path (Fig. 7.17). The landslide debris blocked the path of the river, causing water to back up and flood the land on the upstream side of the landslide.

Restoration of the area has commenced. The river channel was reestablished so that the flood waters could recede from the upstream side of the landslide. Debris was removed to reopen the highway. Vegetation is being established to stabilize and restore the site (Fig. 7.18).

Often evidence of past landslides is visible in the landscape, both as the scars where material has been removed and as mounds or fans where material is deposited at the foot of slopes. Sites of past landslides are likely to move again in future so should be avoided as sites for development of infrastructure. As landslides have occurred in the past in the Oso area, rebuilding housing there is unlikely.

Fig. 7.17 The 2014 Oso mudslide in Washington State, USA. The landslide crossed the river and buried large parts of the road and community in the area. The river was dammed and water backed up some distance flooding the area upstream of the landslide (Photo: United States Geological Service)

Fig. 7.18 The 2014 landslide at Oso in northwest USA. *Left*: the landslide shear plane and debris deposited at the foot of the slope. *Right:* Land re-contouring and re-vegetation of the debris were underway in late 2014

Lahars: Mudflows from Volcanoes

Lahars can be a serious hazard to people living on the lower slopes of volcanoes. Lahars are mudflows formed typically from recently erupted, unconsolidated volcanic material that becomes saturated with water and mobilized. Lahar flows are commonly associated with steep-sided andesite cone volcanoes. Because the source of lahars is high on the slopes of andesite volcanoes, such as Mt. Fuji in Japan, they often flow rapidly (over 50 km/h) down the steep slopes and may extend over 100 km from their source.

Lahars may occur at about the same time as eruptive activity from the volcano. Lahars may also be triggered many years after an eruption when snowmelt or high rainfall causes movement of the relatively unstable material, deposited near the top of the volcano. Mountains such as Taranaki in New Zealand have a "ring plain" built up around the foot of the mountain as a result of repeated lahar flows (Fig. 7.19).

Lahars move fast and carry a mixture of water, sand, and silt. The mineral material gives the flowing liquid a high density which means that it can also carry large boulders and rocks. When the flow finally stops and the water gradually seeps away, the material that is left behind, often in mounds, is an unsorted mixture of sand, gravel, rocks, and anything else that was caught up in the flow. Thus the resulting soil (Fig. 7.20) often contains many large rocks in a matrix of fine material. The mounds make the topography somewhat difficult for intensive land use. However, in New Zealand, pasture for grazing animals is successfully established on many soils on lahar mounds.

Fig. 7.19 The ring plain around Mt. Taranaki in New Zealand is formed as a result of many repeated lahars (mud flows) from the mountain

Fig. 7.20 Soil formed on a lahar mound with many boulders evident. The source of this lahar, Mt. Ruapehu, in New Zealand, is evident in the background. You can see that the lahar has flowed some distance from the mountain

Lahar flows are potentially a serious hazard to cities situated near volcanoes such as Seattle and Tacoma in the USA near the foot of Mt Rainier (Fig. 7.21).

Soil Surface Erosion

Shallow surface erosion removes much of the valuable soil, including the topsoil, which is rich in organic matter that increases available nutrients, improves soil structure, and enhances soil water-holding capacity. Surface erosion is most intense on soil that has no vegetation cover. Soils can blow away in the wind and wash away with water if they do not have a cover of plants to help intercept raindrop impact and roots to bind the soil in place. Wind and water are the main agents of erosion and one severe rainfall or wind event on bare ground can remove soil that may have taken hundreds, or thousands of years to form.

Soil erosion is one of the greatest threats to sustainable long-term food production. Humans eat more wheat, rice, corn and potatoes than all other foods together. To achieve good productivity, wheat, corn, and potato crops require intense weed control, traditionally achieved by plowing. Plowed soil, when bare of vegetation, is vulnerable to erosion by both wind and water. Surface erosion is also a serious threat in areas that are denuded of vegetation due to drought. The problem of drought is often

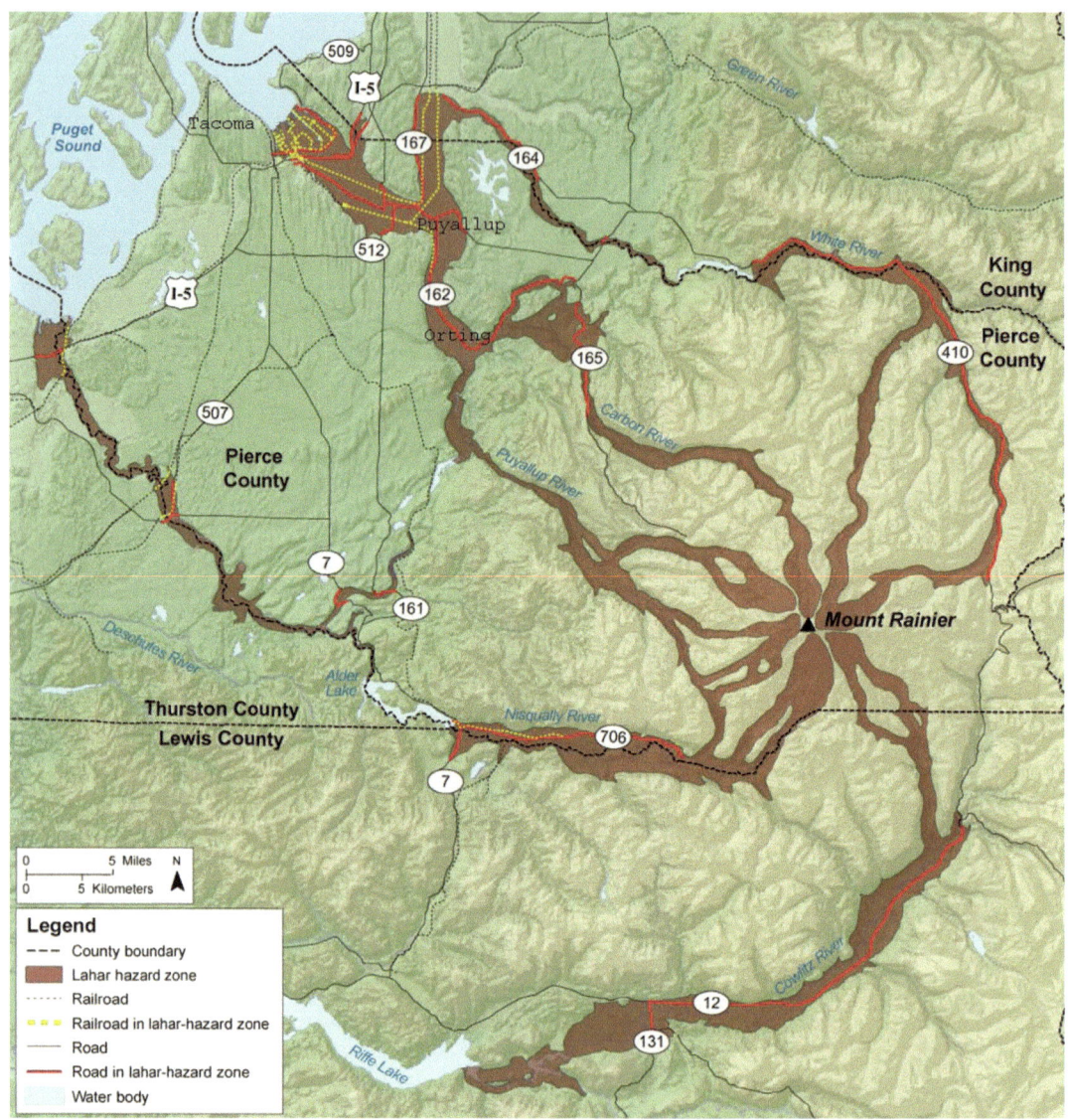

Fig. 7.21 Hazard map of potential lahar flows from Mount Rainier to Puget Sound in the United States. Lahars tend to flow down river valleys but can spread out across the landscape. They have potential to damage low-lying houses and infrastructure such as roads and bridges (Map from United States Geological Survey)

exacerbated in vulnerable semiarid areas by rapidly growing populations who depend on firewood for fuel, thus removing scarce vegetation.

Prolonged drought can lead to removal of vegetation as plants shrivel and die from lack of water and hungry animals eat whatever remains. During droughts, new crops fail to grow in plowed fields leaving the soil prone to erosion. Wind erosion then removes the soil, leaving farmland denuded and causing health hazards in urban areas that may be hundreds of kilometers away (Figs. 7.22 and 7.23). The "Dust Bowl" of the 1930s in the USA is an extreme example of this problem.

Water is the major agent of erosion and removal of material in many regions, whether as a result of heavy downpours, snowmelt, or flood waters that may travel far from the original

Fig. 7.22 Wind erosion occurs periodically in Australia, especially during prolonged drought. *Top*: a plowed field that is vulnerable to wind erosion when the rain fails to come. *Bottom*: a red glow and restricted visibility in Sydney, Australia, is the result of dust blown from bare drought-stricken land, September 2009. The red dust covered cars, buildings, and the grass in this park. Dust from this dust storm even reached New Zealand, over 2000 km away across the Tasman Sea (Photo: Christina Magill)

Fig. 7.23 Wind erosion causes a loss in valuable topsoil. *Top*: plowed soil being picked up and removed by wind. *Bottom*: a US wheat field that has been subjected to wind and water erosion. Normally the topsoil is *dark brown*, but all of the topsoil is missing in the bare area, and the light-colored B horizon is exposed. Very little wheat is growing here as the subsoil is not as productive

Fig. 7.24 Landscape features formed as a result of erosion. *Left*: the exposed roots of a pine tree in Bryce Canyon National Park in Utah, USA, and the extensive eroded backdrop are the result of erosion and removal of soil material by water in periodic intense rainfall events. This erosion is part of a natural erosion process, little influenced by human activities. *Right*: large rill patterns on an eroded gully face that was exposed by quarrying in Thailand

rainfall site. However, removal of soil material can create some interesting and beautiful landscape features (Fig. 7.24).

Celebrating Resilience, Recovery, and the Landforms Created

For all the damage caused when the Earth moves, the land, ecosystems, and people are generally resilient. Following earthquakes, cities are rebuilt and land is rehabilitated (Fig. 7.25). New soils and ecosystems develop on eroded land. Erosion, while often exacerbated by human activity, is also an ongoing geologic process that shapes our landscapes forming an interesting array of landforms some of which we celebrate below.

Eroded soil materials are often deposited on floodplains in the lower reaches of rivers where some of the most fertile and productive soils subsequently form. Flood-derived soil material is often relatively fertile and provides a reasonable substrate for plant growth. For example, once the material deposited during Cyclone Bola in New Zealand solidified, farmers were quick to rehabilitate the land, installing drainage, removing flood debris, and applying fertilizer and grass seed. Over time more sophisticated infrastructure, such as vineyards, has been established (Fig. 7.26) and soil development has commenced.

Fig. 7.25 Part of Napier city in New Zealand which was rebuilt after an earthquake and subsequent fire destroyed most of the town in 1931. The flat area is all newly uplifted land. The new town was constructed in the "art deco" style that was fashionable in the 1930s giving Napier a distinctive character that is celebrated today

Fig. 7.26 An area where flood materials were deposited by Cyclone Bola in 1988. *Left*: farmers were quick to rehabilitate the newly deposited sediment, removing debris, installing drainage, and fencing, then plowing and sowing seed with fertilizer (1988). *Right*: the same site 26 years later in 2014 shows little evidence of the previous damage with a new vineyard established on the mud that was deposited

Fig. 7.27 Glacially carved landscapes are a feature in many parts of the world. *Top*: Swiss mountain landscapes with clear evidence of glacial features including a carved out lake-filled cirque and moraine deposits. *Bottom Left*: a glacially carved mountain peak called a "horn" in the Canadian Rockies. *Bottom right*: steep-sided, flat bottom valleys carved by glaciers in Fiordland, New Zealand. The steep-sided valleys continue beneath the sea in the far distance, forming a fiord

Spectacular mountain landscapes are formed and shaped by processes of uplift and erosion. At high altitudes or high latitudes, glaciers have often contributed to the erosion and deposition processes (Fig. 7.27).

Smaller, but no less fascinating, landscape features occur as a result of differential erosion where some areas of the landscape are more resistant to erosion than others. "Pinnacle" features form in gravel sediments where a large rock or a somewhat more welded layer protects the underlying material from erosion, while that around it is slowly removed (Fig. 7.28). Inevitably the protective layer is eventually eroded, and the feature will gradually be worn down to the level of the surrounding landscape.

In some soil environments, processes within the soil create cementing effects. If the cementation is stronger in some areas than others, then the landscape may be eroded differentially, leaving interesting landscape features, for example, the pinnacles in the Nambung National Park in Western Australia (Fig. 7.29).

Fig. 7.28 "Pinnacle" features in Switzerland where large rocks have protected the underlying material from erosion. Some of the rocks are precariously attached, and once they fall off, erosion of the feature will accelerate

Fig. 7.29 Pinnacle formations at the Nambung National Park in Australia are an intriguing feature. The calcium carbonate cement is derived from shell material in the coastal sands. The reason for the particular shape of the formations is still a subject of debate

Chapter 8
Soils and Humans

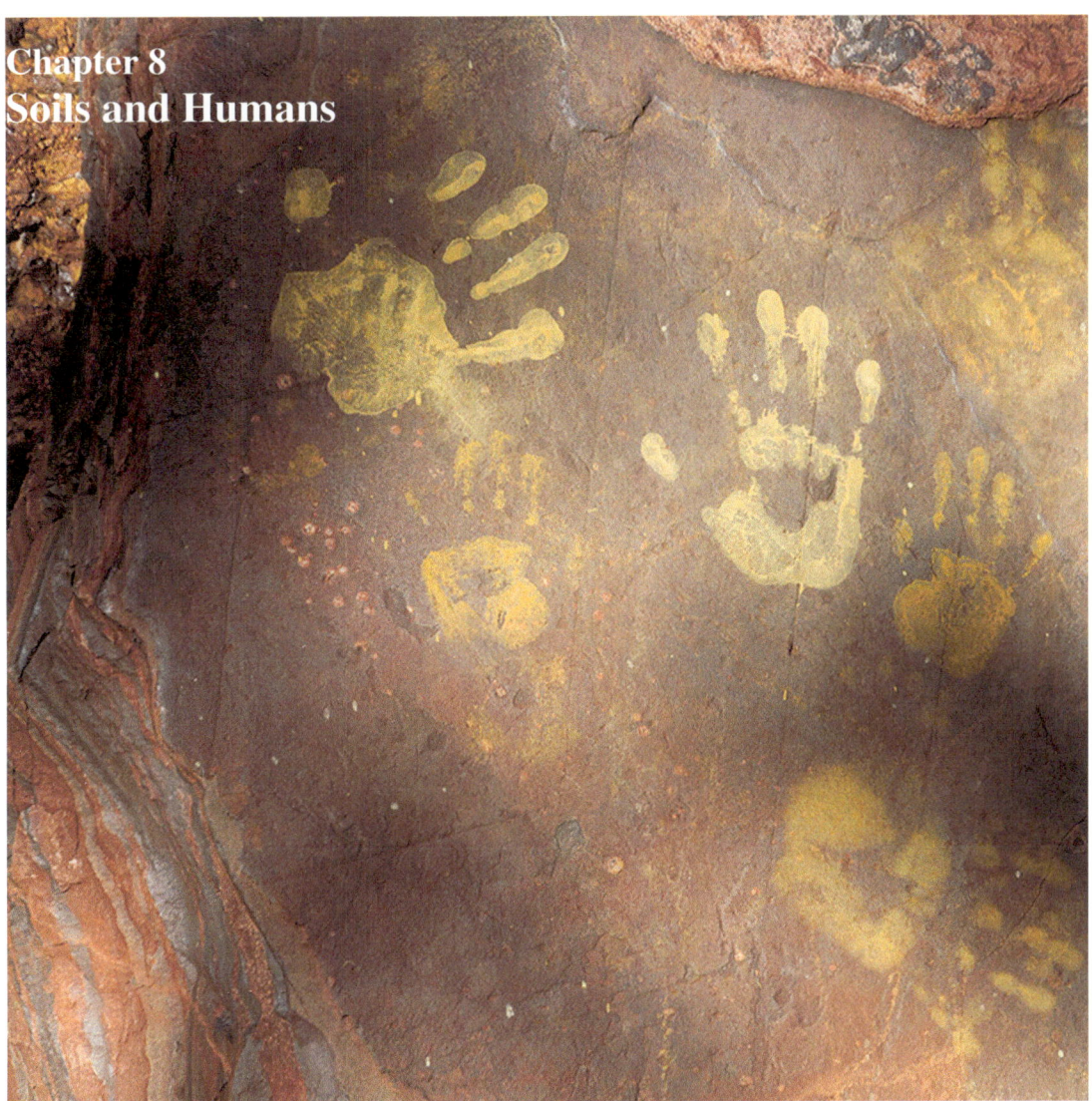

Human handprints formed from clay material on a banded ironstone rock in Western Australia. The rock is some of the oldest on Earth, having formed about 2.5 billion years ago. In contrast, modern humans have inhabited the Earth for only about 200,000 years

© Springer International Publishing Switzerland 2016
M.R. Balks, D. Zabowski, *Celebrating Soil*, DOI 10.1007/978-3-319-32684-9_8

Introduction

"Whatungarongaro te tangata toitu te whenua. People pass on but the land remains".

Maori proverb

The relationship between humans and the soil has, in the past, been close and still is in poorer regions of the world where people's everyday existence relates to deriving their daily food and shelter (Fig. 8.1) from the soil. Our close connection with the soil is recognized in our language; the word "human" is derived from "humus," the organic matter in the soil where life, death, and new life are intertwined.

Fig. 8.1 In many regions, mud or adobe blocks are an important building material. *Top*: a lone worker in Peru toils at making building blocks—they will be used to provide shelter from summer heat and winter cold. *Middle*: house constructed from mud bricks surrounded by bricks drying in the sun. *Bottom*: blocks stacked against a house wall where the flaking plaster reveals the underlying mud brick construction

The connection we feel to the natural world is evident in the arts and beliefs of our earliest cultures. All cultures have long-established links to the soil, sometimes expressed through earth mother goddesses or creation stories where the first people are somehow formed from the soil. We are all aware that following death our bodies will eventually return to become part of the soil again.

In the twenty-first century, with millions living in cities where many people are not directly involved in growing their own food, the direct link between humans and the soil has become more tenuous. In some parts of the world, the role of people who till the soil is seen as lowly and undesirable as work on the land is often harsh and poorly paid. Yet people still find ways to connect with earth. Activities such as gardening are gaining in popularity even in the most crowded cities.

Growing scientific understanding of the importance of soil, along with growing environmental awareness of the impacts of increasing populations and changing technology, is leading to changes in how we regard soil. In the twenty-first century, global culture is gradually changing and requirements to provide for sustainable land management are becoming a law in many countries.

Soils in Myths, Legends, and Traditions

The creation myths of many cultures from widely dispersed corners of the planet have earth mother goddesses and stories related to the formation of humans from soil. Gaia, the earth goddess of ancient Greeks, produced many offspring, some fathered by the sky (Uranus), from which the gods and the earth were populated. Gaia has had a reawakening and a variety of reinterpretations following James Lovelock's publication of the Gaia hypothesis in 1979. Tellus or Terra Mater (Mother Earth) was the Roman equivalent of Gaia.

On the opposite side of the planet from ancient Greece, the New Zealand Maori creation legend has striking similarities to the Gaia story. Offspring of the earth mother, Papatuanuku, and the sky father, Ranginui, include the gods of forests, sea, storms, food, and humans. The subsequent children of the gods include humans, birds, fish, reptiles, and all other living creatures (Fig. 8.2).

In the legends of some cultures, humans are formed more directly from the earth. For instance, in the story from the Hopi people of northern Arizona, a spider-woman made humans using yellow, red, white, and black earth mixed with her saliva. In the Judeo-Christian tradition, the creation of the first human, Adam (the Hebrew word for man and related to Adamah—the Hebrew word for clay or earth), is described in the Old Testament of the Christian Bible, in Genesis 2:4(7) which states, "then the Lord God formed man from the dust of the ground, and breathed into his nostrils the breath of life; and the man became a living being." The Muslim Quran similarly describes Adam as having been formed by Allah from clay, earth, sand, and water. Eve's name means "living"; thus, together Adam and Eve represent soil and life.

Much reverence is often associated with the soil, both as the origin and/or destination of humans and as the earth goddess and the source of food that is vital to immediate survival. The Akan people of West Africa refer to their earth goddess as Asase Yaa. When a member of the Akan tribe wants to prove his credibility, he touches his lips to the soil. Another tradition holds that because Thursday is reserved as Asase Yaa's day, the Akan generally refrain from tilling the land on that day.

Fig. 8.2 A Maori
carving depicting an
important ancestor.
Maori whakapapa, or
genealogy, traces
ancestry all the way
back to the earth
goddess Papatuanuku

Ancient Earthworks in the United Kingdom

Stone circles, henges (ditches and adjacent banks), and barrows (burial mounds) scattered across the landscape of Great Britain provide evidence of past human activities (Figs. 8.3 and 8.4). There are estimated to be more than 1300 stone circles and 10,000 barrows, in the UK, established about 2000–4000 years ago. Construction of these sites involved extensive soil excavation and displacement. Modern archaeologists carefully dig and sieve through the soils at ancient sites to seek clues to the lives and culture of the peoples who lived there.

Barrows, mounds up to 4–5 m high and of varying length, are relatively common in the landscape in Britain and Europe (Figs. 8.4 and 8.5). They are usually formed over a burial chamber in which the remains of a number of people may be interred. Most were constructed in the Neolithic era about 2000–5000 years ago.

One of the largest and best preserved henges is that at Avebury in Wiltshire. Constructed in Neolithic times, about 5000 years ago, the large circular henge encloses an area of about 12 ha with a ditch that would have originally been about 20 m wide and up to 9 m deep (Fig. 8.6). An estimated 200,000 tonnes of chalk rubble was excavated from the ditch and used to build up the 5 m high adjacent bank.

When first constructed, the white chalk bank would have made a spectacular site in the rolling, largely green landscape. A circle of about 100 large (3–6 m high) stones lie immediately inside the henge. Several smaller stone circles, and occasional lone-standing stones (Fig. 8.7), occur within the main enclosure. The purpose of the site is still the subject of much debate and speculation.

Fig. 8.3 A stone circle in the Lake District of Scotland. A smaller stone circle lies within the larger one. The original purpose or form of the site is not clear now as no doubt smaller, more easily moved stones have been pilfered hundreds of years ago for new building projects and any wooden structures have long since decomposed and disappeared. Just the larger stones remain as a monument to past human endeavor

Fig. 8.4 *Left*: a henge preserved at Arbor Low in England originally enclosed a standing stone circle. The stones were probably pushed over in medieval times when people became concerned about pagan sites. *Right*: a nearby barrow. The low-lying area in the foreground was the source of some of the material on the mound

Fig. 8.5 A barrow lies in the middle distance, with the Heel Stone of Stonehenge in the foreground

Fig. 8.6 The roughly circular henge (ditch and bank) that encloses the Avebury site is over a kilometer long

Stonehenge is of course the most famous of all the ancient British sites (Fig. 8.8). The enormity of the stones, and the distance over which they were transported (some more than 240 km from Pembrokeshire in Wales), is remarkable, as is their alignment with the solstice sunrise and other celestial events. Extensive excavation and soil disturbance would have been undertaken during site construction to provide foundations for the large stones and to construct the associated henge.

It is somewhat intriguing that the site that made the word "henge" part of common language actually has a relatively shallow and unobtrusive henge which is largely overlooked by most visitors (Fig. 8.9).

Fig. 8.7 Some of the remaining standing stones at Avebury. Emplacing the large stones required major excavation, not to mention much effort with rollers, ropes, and levers to move them. In order to remain upright, a significant portion of the stone needs to be buried in the ground

Fig. 8.8 Stonehenge stands tall on the Wiltshire plain. The henge, or ditch, that surrounds the spectacular stones is evident in the foreground

Fig. 8.9 Part of the "henge" that encircles the stones at Stonehenge. Presumably over the centuries it has become infilled and is now far shallower than when it was originally constructed

Soils in the Maori Culture of Aotearoa, New Zealand

The Polynesian people who came to New Zealand brought with them a tradition of horticulture and the crops with which they were familiar. Growing frost-tender tropical crops in the cooler New Zealand environment proved challenging. Only two crops survived to become widely cultivated: a sweet potato, known as kumara, and another root vegetable, the taro. The preparation of ground for kumara and taro crops was an important process carried out by well-organized groups of people. Digging sticks were used to loosen the ground, and wooden or stone tools were used to remove weeds and break up clods of soil.

To maximize warmth for the crop, plants were often planted in mounds formed from addition of gravel and charcoal to the soil. The gravel would have improved drainage, thus helping prevent the plants from rotting in wet soil and enabling the soil to warm up faster in the spring. The stones may have also held heat, reradiating it at night, which would have helped prevent frost damage. The mounds also often contain charcoal and it is thought that burned wood ash was added which would have provided some nutrient enrichment. The mounds were always constructed in neat rows (Fig. 8.10), often in a very regular quincunx pattern (the pattern used on dice for number 5).

Much ceremony and tradition surrounded the growing of kumara as it was a difficult crop to grow in temperate New Zealand conditions. Kumara grounds were "tapu" or "sacred" and trespassers could be punished by death. Fences were erected to provide wind shelter and prevent groundbirds from attacking the crop. Weeds and caterpillars were diligently controlled.

In some regions, as populations grew, large areas were cultivated and the gravel for building the mounds was transported for some distance to the garden site. The low-lying alluvial terraces adjacent

Fig. 8.10 A modern recreation of a traditional Maori kumara (sweet potato) garden. The soil, carefully arranged into mounds, had gravel and charcoal added to it. The young kumara plants are just starting to grow through the tops of the mounds. The elaborately carved building is a traditional storehouse (pataka) in which the crop was kept for winter use

Fig. 8.11 Borrow pits from which gravel was removed by Maori are still evident in the landscape on a low terrace adjacent to the Waikato River in northern New Zealand. The gravel was used to form mounds to grow kumara on a higher, less flood and frost-prone terrace near the pits

to some rivers were important sources of gravel and were quarried extensively leaving a pattern of "borrow pits" visible on the land surface from which gravel was removed (Fig. 8.11).

The land surface is also marked with other evidence of former Maori habitation in many areas of New Zealand. Pits were dug in the ground to provide storage for kumara, ovens (or umu) for cooking, and platforms for houses. Sometimes houses comprised a low roof over a pit dug into the ground. Great defensive works were often constructed around hilltops where a combination of trenches and high wooden palisades provided sites that could be defended from invaders by a relatively small group. Some sites proved to be particularly strong, and the Maori defended them for long periods against the superior weapons of the British, during the New Zealand wars in the nineteenth century, gaining the begrudging respect of the British military.

Red ocher (soil rich in red iron oxide minerals), or kokowai, as it was referred to by the Maori, was sought after for personal ornamentation. Possession of red ocher was a privilege of rank. According to some sources, great chiefs painted their whole bodies with it, while lesser individuals may have just painted it on their cheeks. Captain Cook noted its widespread use. Kokowai was often prepared by roasting the soil material in a fire—this would have converted the more common yellow ocher (the mineral goethite) and an orange mineral (ferrihydrite, often found in seepages on stream banks) to the red-colored mineral called hematite.

The Legacy of the Inca Civilization

The Inca civilization in South America recognized the importance of soil and held it in high regard. According to legend, the location of Cusco, the town at the center of the Inca Empire, was determined by pushing a rod into the ground and identifying the area as a site with deep soil. Samples of soil from the four corners of the vast Inca Empire were ceremoniously placed in the main square in Cusco (Fig. 8.12).

The Inca Empire was also noted for exceptional stonework. Stone-surfaced, well-graded roads extended throughout the Inca Empire, and important buildings were constructed with precision joints in huge blocks of stone (Fig. 8.13). The Inca constructed terraces on steep hillsides to facilitate agricultural production. The excellent engineering included drainage installed behind the retaining walls and gravity-fed irrigation systems to deliver water to the terraces. The stonework has stood the test of time with much of it still intact in spite of several hundred years of relative neglect. To facilitate crop production, apparently soil was carried from the fertile alluvial river floodplains to fill in the flat areas on terraces constructed high-up steep mountainsides.

A fascinating site east of Cusco is a circle of terraces which archaeologists have concluded is an agricultural experimental site (Fig. 8.14). Within the terraces are soils that have been brought from a range of sites across the vast Inca Empire, and within the site there are seeds of many varieties of food plants. The circular shape and partial shading of the adjacent cliff provided a wide range of

Fig. 8.12 The modern-day town square in Cusco, Peru, contains soils that were brought from the four corners of the Inca Empire hundreds of years ago

Fig. 8.13 In Peru, superb Inca stonework engineering was used to construct roads and buildings as well as terraces that provided for agriculture while protecting soils on steep hillsides from erosion

Fig. 8.14 An impressively built set of circular terraces east of Cusco in Peru. It is thought that the site may have been used for agricultural experiments as it includes soils moved to the site from far away, seeds of many plant varieties, and a range of microclimates

Fig. 8.15 Traces of agriculture are visible on this steep hillside where exposed soil is readily eroded away by wind and water

microclimates. Like much Inca stonework, the terraces are excellently engineered with drainage behind the walls and in the base of the terraces, topsoil in the upper layers, and irrigation water conducted to the site.

The sustainable terraces of the Inca are not always employed in today's Peru as agriculture has spread beyond the terraced areas. Soil is often plowed (by hand) on steep slopes to grow wheat and corn crops. Inevitably, accelerated erosion ensues (Fig. 8.15).

Middens in British Columbia, Canada

Middens are sites where indigenous people threw waste (often food waste) and it accumulated in deep piles. Middens contain a mix of materials that provide clues to daily life in the past with the oldest material deepest in the soil and the youngest near the top. In the Pacific Northwest, clam shells are a key component of middens, particularly along coastal areas. Clam shells, intimately mixed with sand, soil, and other organic debris, make up a midden. Along the coast of British Columbia, wherever indigenous people lived for long, or returned regularly to harvest seafood, you can find middens (Figs. 8.16 and 8.17). Shell middens are also found in many other coastal areas around the world. Middens form a unique soil because they contain large amounts of calcium carbonate shell material throughout the soil profile. The age of the shell material can be determined using carbon dating giving valuable information about the timing of human habitation in an area.

Fig. 8.16 A midden soil
from the south coast of
British Columbia,
Canada, which contains
extensive clam shells
and organic matter
mixed with beach sands
and gravels. The mixture
of patchy white clam
shells and dark organic
matter are common
features of midden soils.
They are also very sandy

Fig. 8.17 Extremely white shoreline areas indicate long-term accumulations of shells where middens are likely to be found such as this beach in British Columbia

Leaving a Record Using Soil Materials

Indigenous peoples all over the world used minerals that can be found in soils or geologic deposits to draw on rocks and to "paint" their bodies as well as other objects. Hematite is an iron oxide, often known as "red ocher," that occurs in deposits with iron-rich rock types or as a mineral that forms in soils. Old soils may contain abundant hematite. When hematite is finely ground or has formed as fine particles, for example, in rust or highly weathered soils, it is bright red and was used to color carvings and other items and to make pictographs (Fig. 8.18).

In New Zealand, there is an effort to rediscover and preserve traditional crafts using the traditional soil materials. Iron minerals such as hematite (red ocher) and "limonite" (yellow ocher) have traditionally been ground and used to color wooden carvings (Fig. 8.19).

Fig. 8.18 Two examples of pictographs from the British Columbia coast. The pictographs are drawings made using hematite and are often found on rocks near shorelines in British Columbia

Fig. 8.19 Red hematite being ground, for use to color wooden carvings, by a Maori craftsman in New Zealand

The Great Garden Traditions

Many of the great cultural traditions include decorative garden styles which often have many layers of meaning. Traditional gardens often sought to create an enclosed paradise, safe from the heat, dust, noise, and dangers of the outside world. Some of the greatest gardens of the world, such as Louis XIV's garden at Versailles and Peter the Great's summer palace near St, Petersburg in Russia (Fig. 8.20), were developed as pleasure grounds for the rich, but they were also a display of power, wealth, and human domination over nature. As a statement of power, no expense was spared in the creation of the desired effects, regardless of the suitability of the soil. At some sites, major works were undertaken to remove or create hills, lakes, and canals or to import suitable soil. Large numbers of men were often employed on picks, shovels, and wheelbarrows, over extended periods, to build the desired effects.

Chinese and Japanese gardens, while also often developed by the rich and powerful, had strong symbolic significance and were established as places to facilitate quiet contemplation and meditation (Fig. 8.21).

Another garden of importance in Western culture is that of Monet at Giverny in France. Monet's garden was developed as a place where he could experiment with his impressionist painting techniques. Many of the "scenes" such as the reflections of weeping willows in Monet's lily pond

Fig. 8.20 Many of the major gardens of Europe were displays of wealth and power with extensive avenues and gold-plated statuary. *Left*: the extensive garden at Peterhof, Peter the Great's summer palace near St. Petersburg in Russia. *Right*: one of the many fountains and avenues in Louis XIV's famous garden at Versailles in France

Fig. 8.21 Japanese dry gardens are designed for contemplation and contain much symbolism. Raked gravel is often used to represent water. Hamilton Gardens, New Zealand

(Fig. 8.22) were painted multiple times in varying seasons and light conditions. The soil that was removed to create the ponds was used to form higher areas elsewhere in the garden.

The soils upon which the great gardens of the world were established are often largely overlooked by the multitudes that visit them. Many of the sites have been greatly modified and recontoured, to facilitate the grand garden designs that have been imposed, with soil and organic matter imported, if need be, to facilitate good plant growth. The garden at the Château de Villandry, near Tours in France, has perhaps the world's most famous vegetable garden and is a wonderful example of deriving both bountiful and beautiful products from the soil (Fig. 8.23).

Fig. 8.22 Monet's water-lily pond is a well-known garden scene because of its connection to Monet's development of impressionist art

Fig. 8.23 Gardens at the Château de Villandry in France. *Left*: an overview of part of the gardens. *Right*: the detail of the patterns in the extensive vegetable garden illustrates the decorative combined with the productive

"Plaggen" Soils of Europe

In much of Northern Europe, where winters are long and cold, animals have, for centuries, been housed in barns over winter and often overnight during summer to prevent them from straying. For up to 3000 years, in many areas including parts of Germany, Belgium, the Netherlands, and Great Britain, a bedding of turfs of peaty material or soil containing abundant roots from heather, known as "plaggen," was provided for the animals in barns. The remains of the bedding, along with the accumulated animal dung and urine, were regularly removed from the barn and spread over neighboring fields. The buildup of organic material at the soil surface (Fig. 8.24) often reached depths of more than 50 cm and greatly improved the soil productivity. The soils receiving the plaggen material were able to be used continuously for arable cropping for many centuries.

For every hectare of plaggen-enriched soil, where the spent bedding was deposited (referred to as "Eschflur" in Germany), 5–10 ha of source area had the topsoil removed and was thus depleted. The soil and vegetation depletion led to plaggen removal being forbidden in some areas as early as 1798.

In the twenty-first century, the only plaggen removal that is undertaken is to keep some areas impoverished and thus preserve the heather ecosystem that develops in such poor soils. The removal of soil from some areas, and its addition to others, has led to differences in surface elevation of up to about 2 m in some places. In the twenty-first century, it is not only animal manure addition that contributes to the formation of deep dark-colored organic matter-rich topsoils but also a range of other organic wastes from food, energy, and wood production.

Fig. 8.24 Plaggen soils have exceptionally deep topsoils due to additions of manure by people. *Left*: barn in Switzerland from which animal manure and bedding are spread on surrounding land building up the topsoil. *Center* and *right*: plaggen soils from Germany with topsoils over 40 cm thick formed as a result of manure addition. The soil on the *right* is used to grow corn for biogas production, and the biogas digestate is returned to the soil contributing to the organic matter buildup (Photos: Renuka Suddapuli and Sebastian Langheld)

Urban Soils in the Twenty-First Century

Humans have changed soils in the past, and they continue to change them today. Perhaps nowhere is this more evident than in modern cities where most soils have been radically altered from what they were before development. Soils are removed, relocated, mixed, and replaced and rarely look the same after development. All of the five soil-forming factors, climate, organisms, topography, parent materials, and time are likely to be different following urbanization. Whole landscapes may be completely altered to facilitate construction totally changing topography (Fig. 8.25). The microclimate in cities is usually warmer, and less rainwater may move into soils due to impermeable barriers, such as asphalt, concrete, and buildings covering ground surfaces. Often topsoils are removed and new fill materials (sands and gravels from a quarry) are brought on-site thus causing soil formation to begin anew (Fig. 8.26). Vegetation is often completely removed. Where lawns are planted new A horizons may begin to develop quickly but O horizons are lost if a forest was there before development (Fig. 8.27).

Urban development can lead to changes in both physical and chemical soil properties, even where original soils are not replaced. Compaction is a frequent problem of urban soils that often occurs with construction, although it may occur after development from vehicle and foot traffic, especially if vegetation and organic horizons are removed (Fig. 8.28). Even relatively "natural" areas in urban environments such as urban parks show changes to the soil from human activities. Losses or changes to native vegetation will affect the soil profile, soil properties, and the productivity of the soil (Fig. 8.29).

Fig. 8.25 Recontouring of land for urban housing development in New Zealand changes the topography and soil materials on a site. The recontoured site is designed to provide good foundations to support roads and buildings but often leaves the soil less suitable for supporting plant growth

Fig. 8.26 Changes in the soil environment due to urban development. *Left*: a valley bottom soil from Western Washington state prior to urban development. *Right*: the natural soil of this urban valley has been completely replaced with new material that is a more suitable substrate for construction

Fig. 8.27 Soil changes due to conversion from forest to city. *Left*: a forested soil from outside the Seattle urban area showing the soil that would have been present before urbanization. *Right*: housing built over 60 years ago, along with installation of a lawn, shows soil changes. A deeper A horizon has formed under the lawn and the O horizon of the forest is gone

Fig. 8.28 Planting
strips between a
sidewalk and a street
often have a unique soil:
a combination of new
fill that was brought into
the site and
manufactured soil. *Top*:
if planting strips are not
well maintained, they
can become bare and
heavily trampled.
Bottom right: a dense
crust can form at the
surface, due to
compaction, which will
prevent water from
entering the soil and
increase runoff. *Bottom
left*: with careful
planting and
maintenance,
compaction can be
avoided and vegetation
can flourish

Urban soils are subject to many activities that can change over time. Past actions will alter the soil at a site often leaving a record of these changes within the soil. Such ongoing changes make the spatial variability of soils in urban areas very high and cause physical and chemical soil properties to change rapidly and frequently among sites and even within one soil profile (Fig. 8.30).

Fig. 8.29 This urban park near Seattle, USA, has a forest of native trees but has lost almost all of its native understory vegetation. Unfortunately, people have enjoyed this park so much that they have trampled the vegetation and helped erode the O horizon. The loss of the understory and O horizon means that mineral soil can erode too. There is very little A horizon present in the soil profile, and many tree roots are no longer covered by soil

Fig. 8.30 Two soils from the University of Washington Seattle campus in the USA. Even though these soils are not far apart, they each show a record of different past events. *Left*: the dark upper soil is almost all organic matter (compost) that was placed on *top* of the soil after the last road construction project. The *close-up* soil picture shows new gray fill material that was also graded over the site. There is reworked native soil beneath. The very dark material below this contains coal, brick, and other waste buried when this site was a company that supplied lumber, coal, and building materials many decades ago. *Right*: a soil found under a lawn shows a record of additions of new soil materials that raised the soil level in response to past construction over more than 100 years. Some horizons are very compacted and tree roots do not penetrate deeply

Fig. 8.31 Gas Works Park in Seattle, WA, USA, is an example of a contaminated soil on a site that was used to manufacture natural gas from coal and crude oil. Portions of the manufacturing plant remain in what is now a popular park. The contaminated soil has continued to be a problem and is being removed from the park

In some locations chemical contamination, as a result of past activities, is a problem and soils may need to be remediated. For example there are many old gasworks sites in cities world-wide (Fig. 8.31). Site cleanup varies from removing and treating contaminated soils, to adding oxygen, moisture, and nutrients to enhance biodegradation of materials, through to containment and capping.

Celebrating the Productivity of Urban Soils

In Europe, there is a long tradition of allotment gardens where people who live in crowded cities have opportunities to grow food on an allotted space. Allotment gardens remain popular today growing everything from fruit trees and vegetables to flowers. City dwellers often lack space or an appropriate area to grow food, and obtaining a plot in a community garden can solve that problem. Community gardens are becoming more popular as people choose to eat foods they grew themselves, or foods that they know are not treated with pesticides.

There are many benefits to urban community gardens (Fig. 8.32) such as providing locally grown foods that save global transportation costs, keeping more green spaces in cities, and providing urban dwellers with a space where they can get outdoors and watch plants grow and soil be productive! Urban gardens can be educational too, particularly for children who can see where their food comes from and how organic matter is cycled naturally. Organic waste is often composted and reused on-site helping to maintain soil fertility.

Fig. 8.32 Community gardens in Seattle, USA. *Left:* a small plot where individuals or families can rent space to grow their own crops. *Right*: a large community garden with many individual allotment plots evident. There are compost bins for waste organic matter reuse on the site

Fig. 8.33 Backyard gardens grow vegetables, including squash and corn, as well as cheerful flowers

Growing fruits and vegetables within small backyard gardens has remained a popular pastime for urban residents, even though the cost of growing their own food may be more than that of purchasing it. Many urban dwellers enjoy having more direct contact with their environment and enjoy both the productivity and beauty of a home garden (Fig. 8.33).

Fig. 8.34 Roof and vertical gardens on buildings in Paris create attractive outdoor spaces in a crowded city

As cities become more crowded, there is increased enthusiasm for creating green spaces within the built environment. Roof gardens, where soil and plants are imported and established, are gaining in popularity (Fig. 8.34). Green walls where plants are grown against a vertical surface are also appearing in some environments. Because of the constraints of weight on buildings, the soil available to grow plants in has to be kept shallow. Thus, regular watering is needed or plants have to be selected that are highly drought tolerant.

Urban parks, arboretums, and public and private gardens support productive use of urban soils. In many cases, just as with community gardens, the soils in urban parks and gardens are no longer the original soil that formed there, but have been altered or replaced with new material. Nevertheless, soils can be amended and repaired to maintain or increase productivity (Fig. 8.35).

Fig. 8.35 The University of Washington Botanic Gardens/Washington Park Arboretum. *Left*: the soil is generally undisturbed, but the upper 15 cm of the profile is organic matter that has been mixed into the upper soil over the last few decades (each *yellow* and *black bar* of the strap is 10 cm). *Right*: although there is some compaction of the original soil, it is a good foundation for growing plants. A cover of wood chips is being used to prevent weed growth and protect the soil from erosion

Chapter 9
Managing and Caring for Our Soils

Sunflowers, and the bees that pollinate them, are examples of beauty and food derived from the soil. The sunflower was first cultivated in America about 5000 years ago. With care for our soil resources, we hope to continue cultivation of sunflowers, along with a myriad of other plants, for thousands of years to come

© Springer International Publishing Switzerland 2016
M.R. Balks, D. Zabowski, *Celebrating Soil*, DOI 10.1007/978-3-319-32684-9_9

Introduction

"All the seeds, of all the flowers, of all the tomorrows, are here with us today".

Anon

Although soils are incredibly resilient, they are also fragile and can be easily damaged or destroyed. A range of factors limit the long-term vitality of the soil, and to ensure sustainable management of the soil, we need to look as far forward as possible. As the quote above reminds us, all the plants (and other terrestrial lives) that can ever be in the future are dependent on the survival of the seeds and thus the soils, of today. With increasing human knowledge, technology, and population, and thus impacts

Fig. 9.1 The future of our diverse soils is in the hands of everyone on Earth. That future is ensured if young people learn to enjoy, appreciate, and ultimately care for our soil (*Bottom photo*: A. Schipper)

we exert on the Earth, comes increased responsibility for stewardship of our lands and soils. The Earth will remain long after each of us has gone, so whether individuals directly till the soil or not, we are all guardians and beneficiaries of the soil (Fig. 9.1).

In 2012, the human population on Earth reached seven billion. There are now more than twice as many people as the planet was supporting in 1970. Our soil resource is under increasing pressure to provide food for a global population that currently increases by the population of Hong Kong or Madrid about every 4 weeks. As we seek alternatives to fossil fuels, there is also a move to provide energy as biofuels, a product of the soil that is already competing with food production. Our cities are expanding outward over the soil resources on which they depend, and cities also siphon critical irrigation water away from farmlands, reducing soil productivity.

Further challenges for sustainable management of soils for food production include soil loss to erosion, maintenance of soil fertility, accumulation of salts and other contaminants in the soil, and soil compaction. There is a strong, and widely accepted, move to practice sustainable management of soil, water, and other resources.

Having recognized the limits to sustainable use of our soils, we are rapidly developing the knowledge, inclination, and skills, to implement measures to ensure land uses become more sustainable. In this chapter, we review some of the ways in which we can manage soils to both sustain food production and care for the environment.

People on Earth

Unless you live solely on a diet of fish and seaweed, then most of what you eat is ultimately derived from the soil. One of the challenges facing humanity is to feed an increasing population without inflicting irreparable harm on the soil resource. The human population on earth is counted in "billions." For most of us, a "billion" is a fairly incomprehensible number. If you were to count at the rate of 1 number per second, 24 h per day, 7 days per week, it would take about 31 years, 8 months, and 11 days to count to one billion! The human population took all of time until about 1800 to reach one billion. It took about 130 years (to 1930) to reach two billion. It took 30 years (until 1960) to reach three billion, 14 years (until 1974) to reach four billion, 13 years (to 1987) to reach five billion, 12 years (to 1999) to reach six billion, and 13 years (to 2012) to reach seven billion people on Earth (Fig. 9.2). We expect the human population to top out at about nine billion in about 2050 as birth rates gradually decline as more people choose to have smaller families.

The tremendous growth in human population in the twentieth century was only possible with equally rapid increases in food production achieved with increased use of irrigation, fertilizer, mechanization, and pesticides and development of more productive crop varieties. Even so, about one sixth of the population do not get enough to eat for a healthy life.

It is a credit to the resilience of both our soil resource and human capacity for innovation and adaptation that we have managed to continue increasing food production to support growing populations (Fig. 9.3).

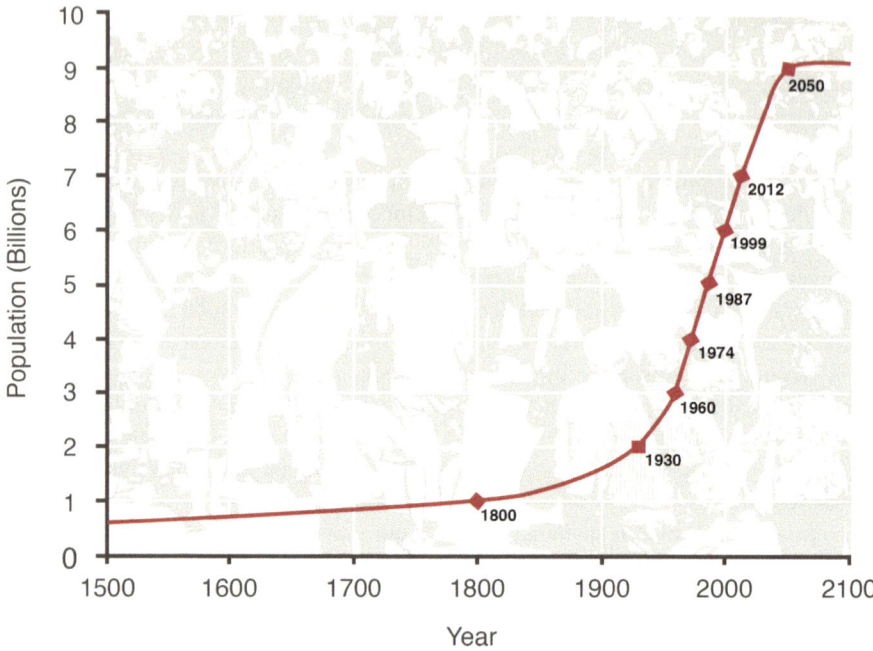

Fig. 9.2 Human population growth since 1500. It is predicted that after the rapid growth in the twentieth century, the human population will reach a maximum of about nine billion in about 2050

Fig. 9.3 As human
population has
increased, the number of
people living in urban
areas has increased

Preventing and Mitigating Soil Erosion

Wherever we observe the soil, it is possible to find evidence of degradation, inflicted as a result of human activity. Plato writing in Greece about 2400 years ago, recognized the effects of erosion caused by deforestation and poor management of hill lands, said;

> "what now remains of the formerly rich land is like the skeleton of a sick man, with all the fat and soft earth having wasted away and only the bare framework remaining." (Plato's dialogue, as quoted in Hillel 1991)

Plato reported that there was loss of arable land and productivity in the hills, while the lowlands were becoming inundated with sediment. He also observed that the loss of soil had caused changes in the hydrology of the area with most rainfall rapidly running off the hills, rather than soaking into the soil and gradually seeping out as springs throughout the year as it had done previously.

The degradation Plato observed in Greece 2400 years ago began in the more recently developed countries, such as New Zealand and the USA, only about 200 years ago. It is still not too late to prevent further damage and to reverse or remedy much of the damage that has occurred. Soils are most prone to erosion by wind or water when stripped of vegetation. Our most important food crops, such as wheat, corn, and potatoes, are traditionally grown in plowed, thus erosion-prone, fields to create a good bed for seed establishment and to control competition from weeds.

We now have alternatives to plowing to maximize plant cover and minimize erosion. Direct drilling (Fig. 9.4), where seeds are planted directly into the soil without plowing, and weeds are controlled using herbicide, is being used more commonly. Leaving crop residues in place, rather than burning them off, protects the soil from erosion and returns organic matter, and some nutrients, to the soil. Plowing around (rather than up and down) the contour and leaving vegetated strips along the contour also help to shorten water flow paths and thus prevent soil erosion (Fig. 9.4).

Strip cropping, where relatively narrow strips of different crops are planted across the contour or across the direction of the prevailing wind, helps to prevent erosion. Under a strip cropping regime, different crops are grown and harvested at different times, so only narrow strips of bare soil are exposed at any one time. Wind protection is provided by the neighboring vegetation, and soil may be moved only a short distance by water before becoming trapped by the neighboring vegetated strip. Windbreaks, such as hedges or lines of trees, help prevent wind erosion and provide shelter for crops thus increasing yields (Fig. 9.5).

Fig. 9.4 Methods to prevent loss of cultivated soil to erosion. *Left*: direct drilling seeds and fertilizer into the soil avoids plowing the soil. *Right*: a fallow field that is accumulating soil moisture. This field has been plowed and will soon be planted with wheat. The plow lines parallel to the contour and the line of vegetation at the foot of the slope help reduce runoff and prevent erosion

Fig. 9.5 Lines of trees form windbreaks to provide shelter for houses and orchards and to help prevent wind erosion in croplands, Canterbury, New Zealand, an area prone to frequent strong winds

Fig. 9.6 Tree planting south of the Gobi Desert in China is an effort to revegetate land to prevent erosion in the headwaters of the Huang He (Yellow River)

Much work is being undertaken to revegetate areas that have been stripped of vegetation and are subject to erosion. For example, large areas of tree planting are being undertaken in some parts of China (Fig. 9.6).

While it would be preferable to confine plowing of soil to flat to gently sloping areas, population and social pressures in some parts of the world mean that food production extends onto steep lands

Fig. 9.7 Crops growing on steep hillsides. *Left*: in northern Thailand plowing and plantings of vetiver grass (a grass species often used for erosion control) along the contour are being undertaken to minimize erosion losses. *Right*: terraces at Machu Picchu in Peru, created with stone walls and filled with soil from the valley below, have provided a stable productive surface for hundreds of years

Fig. 9.8 Trees planted to stabilize landslides while still allowing pasture production for grazing animals in New Zealand. *Left*: established trees and pasture. *Right*: newly planted poplars will help stabilize and prevent expansion of a recent landslide

(Fig. 9.7). Terracing has been used for thousands of years in many regions of the world to enable productive use of steep hillsides. However, terrace construction requires major input of resources, and without ongoing maintenance, soil can be rapidly eroded from damaged terraces. Alternatively, lower-input methods are being developed that may be more achievable for the small farmer. For instance, in the steep hill country in northern Thailand, farmers are encouraged to plow along the contour and leave bands of thick vegetation, such as grass, to prevent runoff and soil loss.

In landslide-prone regions of New Zealand, poplar or willow trees are often planted to help stabilize hill slopes in farmland (Fig. 9.8). Allowing the land to revert to native forest may be the best option in severely eroding areas. In some areas, plantation forestry can give an economic return while providing protection against landslide events.

While erosion is always detrimental, some soils can be surprisingly resilient, and careful management can lead to recovery of an eroded soil's ability to support plant growth. There has been tremendous recovery following the severe landslides that resulted from Cyclone Bola in northern New Zealand in 1988 (discussed in Chap. 7). Twenty-six years later, plantation forests have been established on some of the most vulnerable areas, and pasture has been successfully reestablished on

Fig. 9.9 Landslide damage and recovery in New Zealand. *Left*: damage caused by Cyclone Bola in New Zealand in 1988. *Right*: the same sites (though from slightly different angles) 26 years later in 2014. A plantation forest of fast-growing radiata pine has been established on the most severely eroded areas (*Arrows* point to the same sites on photo pairs)

gentler slopes (Fig. 9.9). On many eroded sites in this environment, within about 20 years, pasture production recovers to 60–80 % of that on un-eroded sites. However, research has shown that there is little further increase in pasture productivity 80 years after landslides in sites like these and plant growth is not likely to fully recover to that of un-eroded sites on human timescales. Similarly, forest productivity has been shown to be about 15 % lower on eroded than un-eroded soils in New Zealand.

In many regions of the world, land use is becoming better matched to land capability with protection forestry on steep slopes, production forestry on moderately steep lands, pasture on lower slopes, and intensive cropping confined to the flatter areas where erosion of cultivated soils is low.

Maintaining and Enhancing Soil Fertility

Removing a crop from a site removes nutrients that were in the soil (Fig. 9.10). It is essential for ongoing productivity that the nutrients removed in crops are replaced, otherwise we are unsustainably mining our soil. Humans who lived a hunter-gatherer lifestyle, or undertook early agriculture, were dependent on the produce from their immediate environment, and nutrients were largely returned to the same environment. Shifting agriculture, crop rotation, and long periods of fallow allowed soils to recover from deforestation and cultivation, giving a reasonably sustainable agricultural system.

Fig. 9.10 Harvesting a crop removes nutrients from the soil. For a sustainable soil system, we must find ways to replace the harvested nutrients. *Left*: hay harvest in Germany. *Right*: women harvesting potatoes in Peru

However, the development of agriculture and organized societies allowed food production to greatly increase, and thus, human populations also began to increase. Food products started to be traded and transported away from the immediate area of their production, thus also removing valuable nutrients.

With the rapid growth in human population and food demand in the twentieth century, it became impractical to leave soils to naturally recover for many years; thus, nutrients had to be replenished. Nutrients can be replaced by adding organic materials including composts, animal manures, wastewater, and biosolids. Nitrogen-fixing plants (such as peas, beans, clover, and alfalfa) can be used to increase the soil nitrogen. However, it is not practically or economically possible to replace all the nutrients removed from the major agricultural areas, such as the Australian wheat belt or the Midwest US corn belt, with composts and animal manures. Thus, manufactured fertilizers are used to maintain food production at levels that keep basic foods affordable (Fig. 9.11). When applied with care, and by matching fertilizer addition to crop needs, fertilizers such as superphosphate and urea increase crop production and have been a key factor in helping food production keep pace with human demand in the latter half of the twentieth century.

There are, however, hazards associated with fertilizer use. If nutrients, particularly nitrogen and phosphorus, from fertilizer get into streams or lakes, they boost the growth of plants and algae in the water. Excess growth causes a decline in water clarity and quality. Eventually, the plants and algae die and microbes decompose the dead material. The microbes may use up all the available oxygen in the water, making the water uninhabitable for most fish species. Careful targeting of the amount, and timing, of fertilizer application can meet crop nutrient requirements without providing excess nutrients that could move into waterways. The riparian margins of streams are increasingly being protected from cultivation, fenced from livestock, and planted in permanent vegetation (Fig. 9.12). The vegetated buffer traps sediment, takes up nutrients, and slows runoff, thus protecting water quality in streams.

Urban centers and many industries produce large volumes of wastewater. Even after sewage treatment, wastewaters often contain plant nutrients such as nitrogen and phosphorus. If the effluent is discharged into waterways, it can cause adverse effects on water quality. An alternative is the irrigation of effluents onto land. If managed carefully, irrigating wastewaters to land can provide increased crop growth as it supplies both water and nutrients to the crop effectively recycling, rather than wasting, them (Fig. 9.13).

Fig. 9.11 Helicopters provide a means of quickly and accurately targeting fertilizer application to the area where it is needed while avoiding adjacent forests or waterways

Fig. 9.12 Riparian planting along a stream margin to protect the waterway from sediment and nutrient runoff on a New Zealand farm

Fig. 9.13 Irrigation of effluent onto land provides both nutrients and water for crop growth. *Left*: treated sewage effluent is irrigated onto a plantation forest near Rotorua, New Zealand. Avoiding direct discharge of city effluent to Lake
Rotorua has led to a marked improvement in the lake water quality over the last 20 years. *Right*: irrigation of effluent to
pasture near Taupo, New Zealand. The pasture grows well and is harvested to remove nitrogen from the site to maximize
the effluent load that can be irrigated without causing adverse effects on the groundwater

Prevention, Mitigation, and Management of Salt Accumulation in Soils

Salt accumulation in soil is a problem in dry regions (as discussed in Chap. 3). Salts may accumulate
in soil as a result of poorly managed irrigation, irrigation with salty water, or where salt-rich groundwater rises to within a meter or two of the ground surface. To maintain soil productivity in drier
regions, it is vital to understand how and why salt accumulates and the measures needed to prevent or
remedy salt buildup.

When developing an irrigation scheme, it is important to prevent salts from accumulating in soil.
Installing drainage and ensuring that the water source is not too salty are important strategies. Careful
control of the amount of water that is applied is important. Enough irrigation water must be added to
ensure that the water passes through the soil washing salts out into the drainage system. However, care
needs to be taken not to add too much water in order to avoid downstream soil waterlogging and water
wastage. Sodium commonly occurs in salty soils and is particularly damaging to soil structure and
plant health. Sodium can be displaced from soil by adding calcium or magnesium.

In some regions of Australia, removal of large trees to facilitate agriculture has led to salinization,
in areas with no irrigation (often referred to as dryland salinity). Without trees, there is less uptake of
water from deep in the ground, and water tables have risen to within a meter or two of the soil surface.
Over five million hectares in Australia are considered to be at risk of, or suffering from, dryland salinity. Tree planting has been undertaken to reestablish deeper groundwater removal in some areas.
Changes in catchment hydrology, as a result of activities such as irrigation or urban development, can
lead to rising groundwater and salt accumulation in the soil. The effects may only become evident
many kilometers from the source of the changes. One such example is an area in Queensland where
upstream urban development impacted water tables leading to salt accumulation and the death of
vegetation including long-established eucalyptus trees. A program to add gypsum (calcium sulfate) to
displace sodium from the soils, and to plant trees to lower the groundwater level, is proving successful. The soils are gradually recovering and vegetation is becoming reestablished (Fig. 9.14).

Fig. 9.14 A site where salt accumulated following upstream urban development in Queensland, Australia. *Top left*: a small creek that became impacted by salt with established trees that subsequently died. *Top right*: the salty soil is inhospitable for most plants. *Bottom*: two stands of trees successfully established on formerly bare, salty areas. Gypsum was added to displace sodium from the soil. The trees help lower the water table

Urban Expansion

As human populations have grown, cities have greatly expanded in area. Large areas of land are also being used to support improved road networks in many regions. Often when land is developed for urban, industrial, or road use, it is covered with concrete, asphalt, or other impermeable surfaces and the productive capacity of the soil is lost.

Many cities have expanded rapidly over the last 50 years. One of the faster-growing cities in the USA has been Las Vegas (Fig. 9.15). The desert land of Nevada is not of great value for productive agriculture; thus, there has been little constraint to expansion and the city has tended to mainly expand outward, rather than upward. The population of Las Vegas has grown from 172,000 in 1964 to 1.9 million in 2010 and it continues to grow.

Growth of cities is likely to continue as population and personal wealth increase and people migrate from rural areas. There is increasing pressure in many regions to consolidate cities within their existing footprint to prevent further spread onto productive land. That means increasing population density within cities and building increasingly tall buildings.

Protection of the soil resource from city expansion requires strong government regulations such as zoning rules, as the economic return from land used for food production rarely exceeds that for urban development. Flat land is the most valuable land for sustainable food production, and it is also the

Fig. 9.15 Landsat satellite images of Las Vegas in Nevada, USA, illustrate the rapid expansion of the city. *Top*: 1972. *Center*: 1992. *Bottom*: 2013 (Images sourced from the United States Geological Survey)

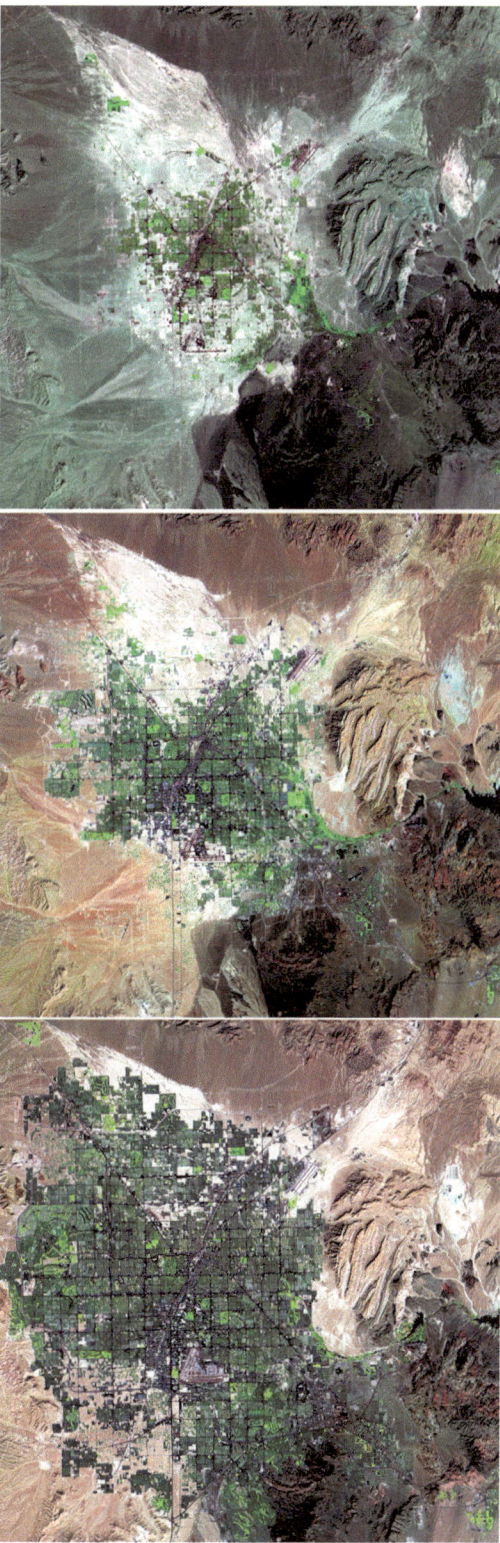

most suitable for urban development. Regulating to encourage urban development onto soils that are of lower productive capacity, such as hill lands, is one way to reduce urban impacts on the most productive soils.

Prevention, Mitigation, and Management of Soil Compaction

Soil compaction damages soil structure and reduces soil porosity. Thus, compaction reduces the air and water that a soil can hold and make available for plants. Compaction can also form hard layers within the soil that become a barrier to plant roots and water moving downward in the soil. The result is reduced plant productivity and sometimes increased vulnerability to surface erosion. Pedestrian, vehicle, or animal traffic can cause compaction, especially when the soil is wet (Fig. 9.16).

Plowing the soil breaks up near-surface compaction, as the soil is worked to form a fine seedbed (Fig. 9.17). However, often a compacted layer will form at the bottom of the plowed layer. Such a compacted layer is referred to as a "plow pan."

Soil compaction slows water infiltration, and in heavy rain, the soil can become saturated above a compacted layer. On sloping ground, severe surface erosion may result as the water-saturated soil moves downslope (Fig. 9.18).

Fig. 9.16 Heavy cattle trampling on wet soil can cause severe damage. The pasture and soil structure of the surface soil have been destroyed. The 10–20 cm deep soil layer is strongly compacted. Sites such as this take many months to recover from the damage, thus greatly reducing the short-term pasture production (Photo: Karsten Zegwaard)

Fig. 9.17 Plowing soil can break up compacted layers near the soil surface. However, plowing leaves soil vulnerable to soil erosion and can cause a compacted plow pan to form at the base of the plowed layer. It is important to plow when the soil is neither too wet nor too dry to minimize potential damage

Fig. 9.18 Severe sheet erosion on a sloping soil with a compacted layer about 10 cm beneath the soil surface. This site, used for grazing cows, was undergoing pasture renewal and improvement. To help prevent erosion, the previous vegetation had been sprayed and new grass seed was direct drilled. After the old grass died off, but before the new grass was established, a heavy rainstorm saturated the top 10 cm of soil and much of it washed downslope. The underlying compacted layer prevented both the water and plant roots from penetrating deeper into the soil. The compaction at this site was most likely to have been the result of animal trampling when the soils were wet

Fig. 9.19 Vehicle tracks visible on a wheat field in Australia. Tractors that travel over crops are set to always travel the same lines in the field to prevent soil compaction over the rest of the land area

There is much that can be done to prevent compaction, and, as farmers have become aware of the production decline caused by compaction, prevention measures are now widely used. When soils are wet, cattle are often grazed for only a short time. The animals are then moved to "standing pads" where they can rest, without damaging the soil, and effluent discharges can be managed.

In cropping areas, tractors often have wide tires to spread the load, thus reducing impact on the soil surface. Vehicle passes are often kept to a minimum, and increasingly, GPS is being used to ensure that vehicles always stay on the same tracks (sometimes called tramlines) so that only a small portion of the area becomes compacted and the rest is not subjected to any vehicle traffic (Fig. 9.19). Compacted soils will gradually recover providing the cause of the compaction ceases. Where earthworms are active, they speed up recovery by mixing, aerating, and reaggregating the soil. Plowing and deep ripping can also be used to break up compacted soil layers.

Prevention, Mitigation, and Management of Contaminants in Soils

Contaminants have accumulated in soils from a range of sources as a result of human activities. Industrial discharges (Fig. 9.20), pesticides, leaded fuel from vehicle exhausts, fertilizers, and nuclear fallout have all contributed contaminants to soil.

Historically, many contaminants accumulated in soil when the adverse effects of activities were unforeseen. Understanding has now greatly increased, and there has been a lot of progress, in many countries, both to remediate the effects of past damage and to prevent new harm from occurring. For

Fig. 9.20 Industrial discharges can spread contaminants over large areas. Much is now being done to prevent discharge of harmful substances to the environment through development and enforcement of environmental protection laws

example, lead has been phased out of petrol, and thus, a major environmental contaminant has been controlled. For dangerously contaminated soils, such as old industrial sites, options can include removing the contaminated soil to a contained site such as a landfill or capping and containing the site, for example, by using the land for car parks where the site is sealed by concrete or asphalt. In some regions, restrictions are placed on contaminated sites that limit the land uses that may be undertaken, or the types of crops grown, to prevent contaminants from entering the food chain.

Cadmium, like other elements, occurs naturally in soils, but cadmium concentrations are slowly increasing in many soils as a result of super phosphate fertilizer use. Much of the phosphorus fertilizer that has been used is derived from guano (bird droppings that have accumulated over thousands of years) (Fig. 9.21). The guano is not only rich in phosphorous but also in a range of other elements including cadmium, fluorine, and uranium. While cadmium is not yet at a dangerous level in most soils, concerns have been raised and research is seeking ways to prevent the cadmium concentration from reaching hazardous levels. Changing to fertilizer sources with lower levels of cadmium and finding ways to reduce cadmium uptake by plants are two possible options.

Copper and arsenic have been widely used to control fungal and bacterial infections in orchards and have accumulated in the soil where they impact beneficial soil microbes and potentially enter the food chain. We can use alternatives to copper sprays on fruit trees to prevent further accumulation of copper from occurring in soil, and breeding pest-resistant plants reduces the need for pesticide use. Some plants, including sunflowers and willows, will preferentially take up and accumulate metals and thus can be used to remove metals from the soil. The metal-accumulating plants can be harvested and the metals extracted for reuse, or the plants can be disposed of in an environmentally acceptable manner.

Fig. 9.21 A gannet colony. The bird droppings, or guano, that were deposited in large seabird colonies of the past provide a source for phosphate fertilizers. However, the bird droppings also contain a range of less desirable elements such as cadmium and arsenic

Sites of industrial activities and military zones are sources of a range of organic contaminants, such as hydrocarbons, and are being remediated using a variety of techniques. At some sites, it is possible to enhance biodegradation of an organic chemical and thus clean up the site (Fig. 9.22). In some instances, removal of the contaminated material to a controlled contained landfill may be the best option.

DDT was widely used to kill insect pests including malaria-carrying mosquitos. DDT was found to accumulate in soil and bioaccumulate up the food chain and has been detected worldwide. DDT in high concentrations can have a number of adverse effects on humans including a possible cause of cancer and reproductive failure. Weak eggshells, due to DDT accumulation, have been identified as leading to declines in some bird populations, particularly of birds of prey (Fig. 9.23). The use of DDT was banished in many countries in the 1970s and 1980s, and remaining residuals in soils are slowly declining as the DDT gradually biodegrades. Bird populations, such as osprey and eagles in the USA and starlings (which eat insects that were targeted by DDT) in New Zealand, have markedly increased since DDT use was banned in the 1970s. In some countries, soils must be tested to ensure there are no high DDT residues before products from the farm, such as milk, can be accepted by processing companies.

Improved pesticides have now been developed that are more targeted in their use and biodegrade rapidly so they do not accumulate in soils. We are now aware of the potential for adverse effects on the soil and wider environment, and new methods are being developed to avoid the need for pesticide use, such as breeding disease-resistant plants.

Fig. 9.22 An example of a contaminated site cleanup. *Left*: an abandoned industrial site in New Zealand with extensive hydrocarbon contamination. *Right*: biodegradation was enhanced by adding nutrients and oxygen to the material until a safe inert product was formed that could be incorporated into the soil at the site. The site is now a protected "historic place" due to its industrial history

Fig. 9.23 A condor in Peru—birds of prey, high in the food chain, were vulnerable to the cumulative effects of DDT

Precious Water: A Key to Soil Productivity

Irrigated agriculture accounts for about 70 % of global water use, and irrigation water is critical to maintaining soil productivity in many regions. Clean freshwater (Fig. 9.24) is a vital resource for all terrestrial life. Earth is known as the blue planet, due to the huge water resource. However, about 97.5 % of Earth's water is in the sea (Fig. 9.25) where it is too salty to use for drinking water or for irrigation. Of the remaining 2.5 % of water that is "freshwater," about 75 % is ice (mainly in Antarctica), 22 % is hidden from sight in groundwater, and only about 1 % of freshwater (or just 0.025 % of all the water on the planet) is held within rivers, lakes, the atmosphere, and the biosphere.

Water is generally not created or destroyed, but just moved around the planet via the hydrological cycle. The process of evaporation of water into the atmosphere and its return as rain or snowfall is a process of distillation that effectively cleanses and renews our water resource. Water moving through the soil is effectively filtered, often also improving water quality. However, water availability is the

Fig. 9.24 Clear freshwater is one of our planet's greatest assets

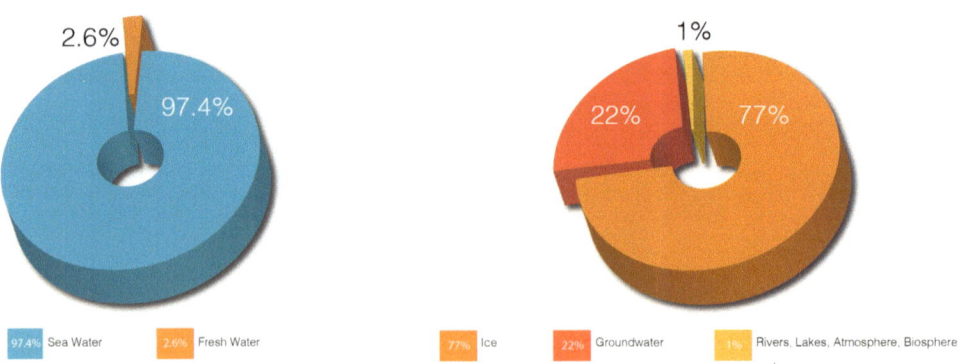

Fig. 9.25 Most of the water on earth is seawater, and only a very small portion is freshwater that is readily available for use. *Left*: global distribution of water. *Right*: distribution of freshwater

greatest limitation to plant growth. Of all our resources, water is coming under the greatest stress from the increase in demand that has come with increasing human populations and the increasing sophistication of our society.

Some irrigation water comes from groundwater that was accumulated in a wetter climate thousands of years ago. Often, water is being extracted from the groundwater at a faster rate than it is being recharged and thus the water is effectively being mined. The Midwest USA is such a region. The water table in the Ogallala aquifer (a 450,000 km² groundwater reservoir that underlies parts of South Dakota, Texas, Nebraska, New Mexico, Wyoming, Colorado, Kansas, and Oklahoma) is dropping

Fig. 9.26 Water is the main limitation to plant productivity in many parts of the world. The green areas are receiving irrigation, while the adjacent gray-brown areas remain parched and dry. *Top*: high Andes, Peru. *Bottom*: Canterbury, New Zealand

such that some predict that if current practices continue, much of the extractable water will be gone by about 2030. As cities and industries grow, so too does their demand for water. The price that is paid for much agricultural produce means that farmers cannot afford to pay as much as neighboring cities for access to water. Thus, already some areas that were formerly irrigated for food production have been returned to dryland farming, particularly in Texas. The natural recharge in the Ogallala aquifer is such that it would take thousands of years to replace the water that has been removed since large-scale irrigation began in the 1940s.

In dry regions, production is usually considerably lower on land that lacks irrigation than on adjacent irrigated areas (Fig. 9.26). Strategies to prolong the water supply include developing more efficient irrigation systems; planting crops, such as sunflowers, with lower water demand; and reusing treated city or industrial wastewater for irrigation.

Much has, and continues, to be done to improve water-use efficiency. Israel has a mean annual rainfall of only 435 mm and shares its main water source, the River Jordan, with Syria, Jordan, and Lebanon. As a result of water scarcity, Israel has put huge effort into developing efficient water use. Thus, Israel has led the world in developing irrigation systems using micro-sprinklers and drippers to target water specifically to plant root zones and meet crop requirements with minimal water wasted. Israel is developing subsurface drip irrigation to make water use even more efficient and to allow irrigation of recycled wastewater without potential for surface contact with edible crops. Australia also has severe water limitations and is becoming much more efficient in both water use and wastewater reuse.

Drainage to Improve Soil Productivity

Where soils are waterlogged, drainage has been used to greatly improve soil productivity. The Alblasserwaard region of the Netherlands is close to sea level and soil drainage was first installed in the thirteenth century. Drainage led to land subsidence, while sand accumulation in the river bed raised the river level. Eventually, it became necessary to pump water up, over the flood prevention dikes, to maintain the productivity of the land. The famous windmills near Kinderdijk (Fig. 9.27) were built to pump water from local drains up into larger canals and ultimately into the Lek River where dikes (or flood stopbanks) protect adjacent land from inundation.

In Thailand, drainage installation and water management have enabled crops to grow in places where land-use options were limited by high water tables (Fig. 9.28).

Organic soils can be developed and used for productive land uses. However, the peat resource will inevitably gradually be lost as a result. In order to grow pasture or crops on peatlands, it is necessary to drain the soil (Fig. 9.29) and apply fertilizers. Some organic soils tend to be low in nutrients and have a low pH. Thus, large additions of lime and fertilizer nutrients such as phosphorus, nitrogen, and potassium are required for many crops. Drainage leads to oxygen becoming available in the organic material. Microbes then gradually decompose the peat converting it to carbon dioxide and water. The addition of nutrients also benefits the microbes, helping increase the rate of organic matter decomposition. The soil surface gradually subsides until eventually the bottom of the peat layer is reached. Drainage therefore becomes more difficult and expensive over time. If drainage can be maintained, then farming may continue, in the long term, on the underlying mineral material.

While drainage is important for productive agriculture, natural wetlands are special places with their own assemblages of unique plants and animals (Fig. 9.30). Wetlands have a key role in enhancing water quality and storing water, taking the peaks out of flood events and allowing water to move slowly into streams, helping maintain flow during drier episodes.

Fig. 9.27 Cows and horses graze pasture near Kinderdijk in the Netherlands where drainage works commenced about 800 years ago. Pumping water out over the dikes (using windmills from about 1740 until the mid-twentieth century) continues today

Fig. 9.28 *Left*: at this site in Thailand, drains have been excavated and the soil has been mounded up to provide a drier site for fruit trees. *Right*: the soil has a relatively high water table and rock material near the surface. However, with drainage and good husbandry, fruit crops are successful

Fig. 9.29 Drainage in peatlands, accompanied by fertilizer addition, can lead to productive pastures. Keeping the water table as high as possible, as shown in this example, helps slow the rate of subsidence and prolongs the life of the peat soil resource

Fig. 9.30 Wetlands on the margin of a small lake support a unique assemblage of plants, animals, and invertebrates. The wet conditions mean that peat soils will be forming beneath the reed vegetation

Celebrating the Variety, Productivity, and Beauty of Soil

"The colors of the subsoil are as radiant as those of a sunset, and both exist in a world on the edge of darkness."
Peter Singleton

Soils are the hidden foundations of the landscape and society. Beneath a magnificent view, there is a soil providing water, nutrients, oxygen, and anchorage, for the life that grows there. While soil, on first impression, seems a fairly simple material, soils are astounding in their complexity, variety, and resilience. Like snowflakes, no two soils are the same (Fig. 9.31) and soils arguably comprise the most complex ecosystems on Earth.

Caring for the environment means caring for its soil. Without the benevolence of soil, neither we nor the environment will thrive. Food is the first thing most people think of when considering productive soils, but as we have discussed, we get many other things from soils such as clean water, fuel, building materials, fiber, habitat, and medicines. Soils also have an important role in disposing of our wastes and recycling the nutrients that are vital to life on Earth. Soils are also critical to the global cycling of elements that maintain our oceans and landscapes. The true value of soils is often hidden just as soils themselves are hidden.

We need to occasionally pause in our busy lives and celebrate the bountiful myriad of products of the soil. We need to cultivate a sense of gratitude for the soil's harvest: our forests, food, and flowers. Such appreciation can encourage care and good management to continue to produce food for our survival, flowers to lift our spirits, and forests for fuel and shelter, while maintaining the potential for ongoing production in the future (Figs. 9.32).

Fig. 9.31 Soils have tremendous variety in color, texture, and the life they support, all of which are a function of the environmental factors that interact with and impact on soil processes

People who work closely with the land know the value of soils, but may not take time to appreciate their beauty. People who live in urban areas rarely see soils and often fail to acknowledge how critical they are to their daily lives and think of them merely as "dirt." There is a saying among soil scientists that we should "never treat our soils like dirt." The best way to learn and get to know soils is to get out there and get close to the soil. Plant a garden, make a clay sculpture, and enjoy the feel, sight, sound, and smell of the soil in its endless dance with water and life. Celebrating the beauty and productivity of soil may help each of us to recognize that during our brief stay in this life, we all have a role as guardians, to ensure that soil is sustained to support not only our present needs but also "all the flowers of all the tomorrows."

Fig. 9.32 The bounteous products of the soil. Food, forests, and flowers are some of our most valuable riches

Appendices

Appendix 1: Visual Dictionary

A picture is worth a thousand words! Some basic soils, landscape, and geologic terms are defined here using illustrations and explanatory text.

Aggregation

The binding of soil particles into clumps (see *Soil Structure*).

Allophane

A poorly crystalline hydrated aluminosilicate mineral (contains Al, Si, O, and H) that usually forms in well-drained soils with volcanic parent materials, especially soils high in volcanic ash. Allophane typically occurs as spheres that are only a few nanometers in diameter. It does not have an exact chemical formula, but develops within a range of concentrations of silica and water. The soil shown in this photo is rich in allophane; the allophane allows the soil to drain but still retain water within micropores. The allophane can bind phosphorus making it unavailable to plants. Allophanic soils are friable and allow roots to easily penetrate deep into the soil.

© Springer International Publishing Switzerland 2016
M.R. Balks, D. Zabowski, *Celebrating Soil*, DOI 10.1007/978-3-319-32684-9

Alluvium

Layered river or stream deposits. Alluvial deposits often vary between coarser material (carried by faster-moving water with more energy that can carry bigger materials) and finer materials deposited from slower-moving waters. Rocks are typically rounded from abrasion while being moved by water. This photo shows a soil formed in recent alluvial materials, with rounded, coarse flood-borne rocks near the surface and alternating deposits of finer and coarser sands and rocks below.

Andesite

A volcanic rock of grayish-green-brown appearance with some coarse minerals in a matrix of finer minerals. Andesite has a moderate silica content with a mixture of minerals imbedded in a glassy matrix. The andesite shown here contains plagioclase and quartz with some blackish iron-rich minerals such as biotite (black) and chlorite (green). Andesite has a moderate viscosity; thus when andesite erupts from a volcano, it is somewhat explosive but will form thick flows of molten rock.

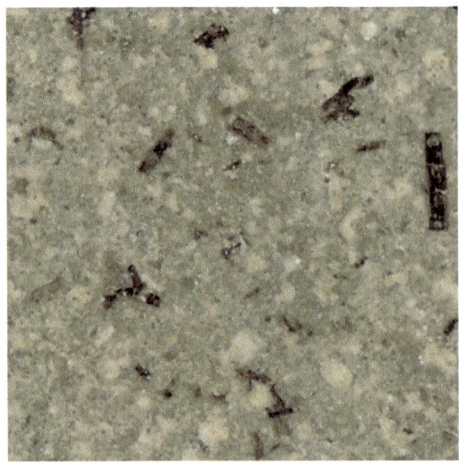

Basalt

A fine-textured, blackish volcanic rock that has a high iron and magnesium content. Basalt has a low viscosity and will produce broad, extensive lava flows. Basalt cools quickly so only small mineral crystals can form thus its fine-textured appearance. Large basalt flows can form large hexagonal columns as it cools such as those shown here.

Calcium Carbonate (Carbonates)

A mineral with a composition of $CaCO_3$. Finely ground or poorly crystalline calcium carbonate is also called lime. Lime (or ground limestone) is often applied to soil to raise pH. This photo shows lime being applied to a garden soil. Other metals such as magnesium can combine with carbonate to form carbonate minerals. Carbonates often accumulate in soils with little leaching.

Clay

Very small particles that are less than 0.002 mm. The term clay is used for any particles of this size but is also used to indicate specific minerals that are flat and layered; these are sometimes called sheets of clay. Clay can form in soils, but some clays can be inherited by a soil from rocks. This scanning electron microscope photo (left) shows a large sand particle that is covered in smaller particles—the small flat particles are sheets of clay. Clay also includes small particles of round or tubular-shaped minerals that form in volcanic soils. On the right are very large stacks of particles of clay called vermiculite (vermiculite is usually too small to see in soil). Stacks of clay are also called books of clay. The scale is inches and centimeters.

Decomposition

The breakdown of organic matter into smaller molecules by soil microbes such as bacteria and fungi. Organisms decompose litter to get the energy stored in organic matter and release carbon dioxide or form humus (see *humus*) during this process. Larger soil organisms help decomposition by breaking organic litter into smaller pieces that are easier for microbes to decay further.

Eluviation

Eluviation is a specific type of leaching: clay particles, dissolved salts, iron, aluminum, or organic matter can move *out* of one soil horizon with water moving down the soil profile. In some cases elements are weathered out of solid particles and become soluble ions that move with soil water; salts are an example of this. Soil water can also move very tiny (colloidal) particles that are suspended in the water out of a horizon. E horizons are formed by leaching of materials. Movement of leached materials into a lower horizon is called illuviation. The left photo shows a soil profile that has developed an eluviated E horizon from the movement of clay to lower in the profile. The right profile shows movement of Fe, Al, and organic matter from a gray E horizon. Note that the B horizon has accumulated these materials, which is illuviation.

Floodplain

Land that is adjacent to a river or stream that periodically floods. Because floodplain soils can be inundated with water, soils in a floodplain can be very different from soils just out of the floodplain. Vegetation can also change rapidly at the boundary (see *riparian*). A black line marks the edge of the stream floodplain in the photo on the left; the entire flat valley bottom is the river floodplain in the photo on the right.

Glacial Outwash

Outwash is similar in appearance to alluvium; the main difference is that glaciers are the source of the river water putting down the rounded rocks and sand in layers. Glacial outwash can fill glaciated valleys making a U-shaped glacially carved valley become flat as it fills with outwash.

Glacial Till

Glaciers can be thousands of meters deep; deep ice is heavy and will compact any loose material under it. As glaciers move, they grind away the rock beneath them and ride on this fragmented material, compacting it at the same time. This picture shows a close-up of compacted basal till that was once underneath a large continental glacier. Note the predominantly fine matrix material with larger rock fragments mixed in. Compacted till like this not only looks like concrete, it is just about as hard and water does not flow through it easily. There are sands and rocks carried in and on top of glacial ice that drop when the ice melts. This is loose material that is called ablation till.

Gravel

Rock or mineral material that is greater than 2 mm in size but less than 75 mm (see *sand*).

Hematite

An iron oxide (Fe_2O_3) that is a rusty red color when it is poorly crystalline or black when very crystalline. Hematite is usually red in soils as it is poorly crystalline and does not form large crystals; it often coats other particles. Hematite forms when there is plenty of oxygen available to combine with iron. This example shows nodules of reddish hematite that formed in a very old soil. The scale is in cm.

Humus

Well-decomposed organic matter that is darkly colored. Plant and animal litter is decayed by numerous organisms (bacteria, fungi, insects, arthropods, etc.) that break down large organic molecules which are then recombined into a new amorphous brownish-black substance that can retain water, provide nutrients, help bind soil particles together, and enhance soil fertility. This photo shows two O horizons: the upper soil is well-decomposed organic matter that is all humus, but the lower soil is less decomposed organic matter that is a lighter brown color and retains fragments of plants that are not completely decomposed and still recognizable.

Illuviation

Leaching of materials *into* a soil horizon. Illuviated horizons have accumulations of materials such as salts, clay, organic matter, iron, and aluminum (see *eluviation*).

Laterite

A soil type usually found in tropical or subtropical areas. Laterite soils form in areas that have both high rainfall that will strongly leach soils and a stable old landscape with parent rocks that are rich in iron. The intense leaching strips away most elements, but iron and aluminum are left behind as oxides giving the entire soil a very reddish-orange color. Decomposition is rapid because of the warm temperatures and abundant moisture, so there is usually a pale A horizon in these soils. Each section of the tape is 10 cm.

Leaching

The movement of soluble or suspending materials from one horizon into another (see *eluviation*) or out of a soil into groundwater.

Mineral

A naturally occurring inorganic or organic solid with a specific chemical composition and arrangement of atoms. Each mineral has a repeating arrangement of atoms in specific organization that gives the mineral a characteristic crystal shape; poorly crystalline minerals may only have this characteristic structure for a few molecules and never form larger crystals. Rocks are composed of minerals. Some common minerals are quartz (SiO_2), feldspar ($KAlSi_3O_8$), and biotite mica ($K(Mg,Fe)_3(Al,Fe)Si_3O_{10}(OH)_2$) shown here from top left to bottom right.

Nutrient

Elements that are essential for life or for an organism or necessary to complete its life cycle. For example, plants need carbon, hydrogen, oxygen, nitrogen, phosphorus, potassium, sulfur, calcium, magnesium, iron, boron manganese, copper, zinc, chloride, molybdenum, and nickel. Some plants need additional elements, such as legumes which need cobalt. Without adequate amounts of a nutrient, a plant cannot grow well as shown in this photo of branches from two cedar seedlings. The left branch is from a seedling growing in soil with plenty of all essential nutrients. The branch on the right is from a seedling growing in soil with all nutrients except nitrogen—this lack of nitrogen stunts the growth of the tree and makes the foliage look yellow or chlorotic (photo: Stan Gessel).

Oxidation

The loss of electrons by an atom or molecule. When oxygen is present in soil, it can accept electrons from other elements—oxygen accepts electrons provided by another element giving the oxygen a negative charge which lets oxygen bind with a positively charged ion. Oxygen can combine with these positively charged ions to form oxides. Rust is an iron oxide (see *hematite*). Oxides of iron and aluminum form when a soil has oxygen. Oxides often precipitate out of solution onto particle surfaces (see *laterite* and *podzol* for soil horizons showing oxidized iron). Reduction is the opposite of oxidation.

Podzol

A type of soil that shows development of a soil profile with eluviation and illuviation of oxides and organics (leaching of soluble materials from one horizon into a lower horizon). These examples show a visible loss of iron from a gray E horizon into a reddish B horizon. With this particular soil, organic acids form in the O horizon and dissolve iron (Fe) and aluminum (Al) from minerals in the E into soil water. The iron and aluminum then move down with the soil water. The iron and aluminum precipitate and accumulate in the B horizon. Organic matter is also accumulating in the upper B horizon of the profile on the left (a much darker B horizon).

Pumice

A glassy volcanic rock that forms from molten rock quickly hardening to glass as it rapidly cools after being ejected by a volcano into the air. It is typically a light color, has a high silica content, and is often so lightweight that it will float in water. Pumice is lightweight because air is trapped inside bubbles of a glass as the rock cools. Scale is in cm.

Reduction

A gain of electrons by an element or molecule. When soil air has no oxygen, it is said to be reduced or anaerobic. Reduction is a chemical reaction where electrons are transferred from an element such as oxygen to iron or another element giving the element a more negative charge. In soils, when there is no oxygen present, iron will be reduced and loose its reddish color making the soil look gray. Very reduced soils are all gray. Soils that are reduced in places will have patches of gray and yellow/orange colors (see *soil mottles*).

Rendzina

A type of soil that has a dark blackish-brown A horizon that is high in calcium and underlain by limestone. The calcium released by weathering of the limestone will bind with *humus* making it very stable; thus, humus can accumulate and the A horizon becomes very dark with a high organic matter content.

Rhyolite

A volcanic rock that has a light color and a high content of silica. Rhyolite has mostly small crystals due to its rapid cooling after eruption, which give it a fine texture with some larger crystals embedded within a glassy matrix. It has a high content of quartz and feldspar (thus a high silica content) with trace amount of dark iron-rich minerals such as biotite. Rhyolitic volcanoes tend to have more explosive eruptions without big flows of molten rock.

Riparian

The land and ecosystem adjacent to a river or stream influenced by the river. Riparian areas typically have wetter soils and vegetation adapted to both wet soils and periodic flooding. Riparian areas are strongly influenced by the river they are adjacent to, with both inputs and outputs of water and sediment. This figure shows deciduous trees (without their leaves) in a riparian area adjacent to a river; the coniferous trees are above the riparian area.

River Terrace

A flat or nearly flat surface that was formerly the alluvial deposit of a river floodplain. Rivers carry sediments that eroded from upstream areas and deposit sediments downstream often creating broad floodplains that fill in valleys. River deposits can accumulate and the river may cut down through these deposits, essentially abandoning old floodplains that are no longer part of the active river channel. The former floodplain surface is called a terrace (risers are the steep slopes that connect terraces). This photo shows four river terraces. This photo shows four river terraces indicated by white arrows. The river has now incised deeply into the alluvial deposits and flows at a much lower elevation than the terraces shown here.

Sand and Gravel

Coarser mineral soil components include sands and gravels (and larger rocks). Sand is 0.05–2 mm in size and is part of soil texture (see *soil texture*). Gravel is a modifier of soil texture and added to a soil texture name if it is present. These photos show typical rounded/blocky/irregularly shaped sand particles on the left and sand with gravel-sized cobbles on the right.

Scoria

A glassy volcanic rock that forms from molten rock quickly hardening to glass as it rapidly cools after being ejected by a volcano into the air. It is typically a dark color with a higher iron content. Similar to pumice, it is often so lightweight that it will float in the water.

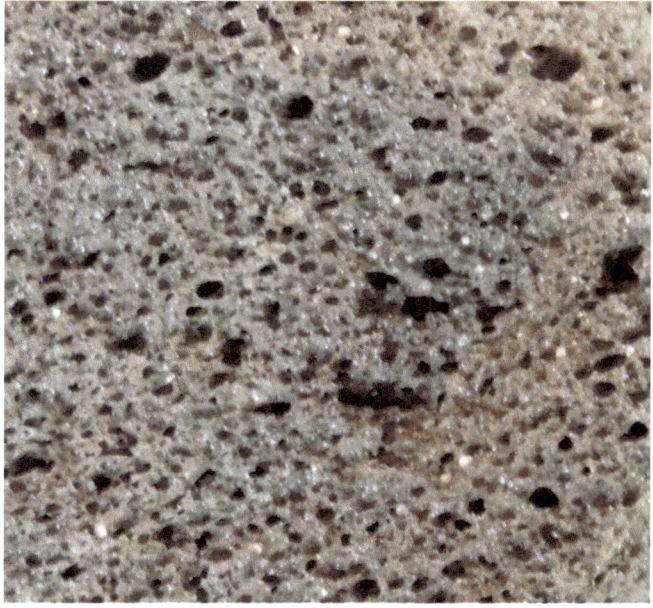

Silt

Silt particles are 0.002–0.05 mm. They are similar in shape to sand particles (blocky, rounded, and irregular) but smaller. This photo shows silt on the left compared to sand on the right. Note that the silt particles are so small; you cannot distinguish individual silt particles without magnification as you can with sand. The scale is in cm.

Soil Horizon

A layer of soil that is different from the soil above or below due to a change in color, texture, composition, aggregation, or some other features. This example is a simple soil profile with only two horizons evident (an A horizon is the dark topsoil and a B horizon in the reddish subsoil). See Chap. 1 for a description of the main types of soil horizons.

Soil Mottles

Soil mottles of color form when part of the soil has pores that become depleted of oxygen and part of the soil has oxygen remaining in the soil air. Uptake of oxygen by roots and microbes lowers soil oxygen levels. Atmospheric air with oxygen will move into the soil (often as water drains out) to replenish the soil air and oxygen. Where there is oxygen, the soil will have a brighter color such as yellow, orange, red, or tan—this is because oxides of iron can form when oxygen is present and they have colors just like rust. Where there is no oxygen, the soil will be gray, bluish, or greenish. These two examples show soil with intense mottling. Both examples are old, clay-rich soil so the pores are small and don't easily get oxygen replenished once roots or microbes use up the oxygen. The more bluish soil has much less oxygen than the reddish soil which is only lacking oxygen in some areas.

Soil pH

A measure of the acidity or alkalinity of a soil. Soil pH is the negative logarithm of the concentration of H^+ ions in a solution of water and soil. If a soil has a pH less than 7, it has a high concentration of H^+ and is acidic. A pH of 7 is neutral, and a pH greater than 7 is alkaline or basic and has few H+ ions. Extreme acidity (pH near 1 or 2) or alkalinity (pH near 14) can prevent vegetation from growing. Forest and wetland soils are usually acidic (pH 4–6). Grassland soils are closer to neutral. Desert soils are usually alkaline. A good pH for an agricultural soil is about 6.5–7.

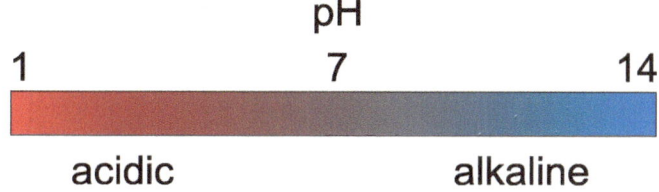

Soil Porosity

About 50 % of soil is open spaces between soil particles, called soil pores. The total amount of pore space a soil has is referred to as soil porosity. In this diagram all the gray forms are solids and the white spaces represent pores. Pores can be filled with either water or air, and the percentage of water or air in a soil will change with wetting and drying. Soils can have large pores that allow rapid movement of air and water or small pores that retain water.

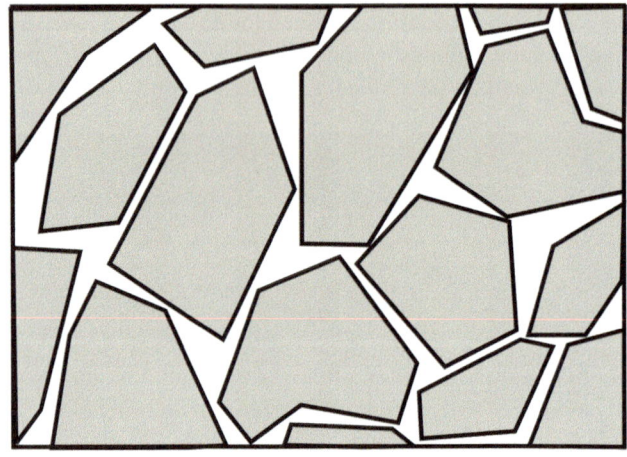

Soil Structure

Soil particles can be aggregated or clumped into various shapes. Examples shown here include (in clockwise order from the upper left) crumb (small rounded), blocky, prismatic (tall and narrow), and platey (flat usually with layers). Having aggregation helps protect soils from erosion and give soil bigger pores (between aggregates) and smaller pores (within aggregates) allowing water retention and movement. Soil can also have all particles cemented together (a massive structure) or have no aggregation (single grained).

Soil Texture

A designation indicating the relative amount of sand, silt, and clay in a soil. The percentage of sand (2–0.5 mm), silt (0.5–0.002 mm), and clay (<0.002 mm) in a soil is plotted using the texture triangle to indicate soil texture. There are typically 12 different soil textures recognized. The soil triangle shows these different textures with the percentage of each particle size on different axes. A loam soil has approximately 40 % sand, 40 % silt, and 20 % clay. This is a great texture for a productive soil. Note that organic matter is not included in soil texture.

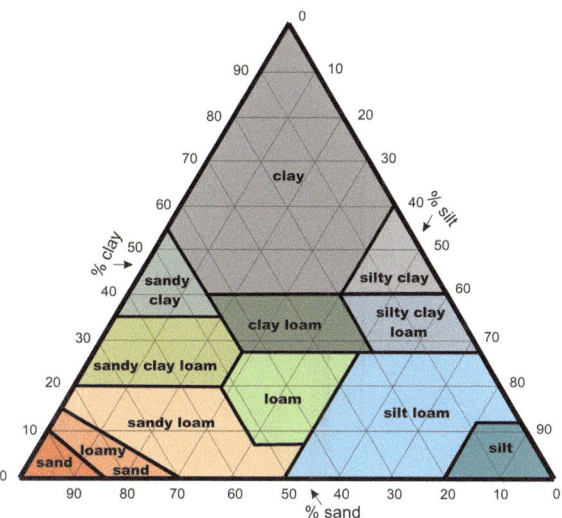

Tephra

A mixture of glassy volcanic ash with fragments of *minerals*. Violent volcanic eruptions will eject small, glassy particles of *volcanic ash*. The ash may be mixed with fragments of mineral crystals. These mineral fragments were in the rocks before the eruption blew the rocks apart. When ash (typically light colored and high in silica) is mixed with fragments of minerals (some of which may be darkly colored), it is called tephra. The tephra shown here is from an eruption of Mt. St. Helens.

Urea

A nitrogen fertilizer. Urea is an organic compound [$CO(NH_2)_2$] that is about 46% nitrogen by weight. As a concentrated source of nitrogen that can be manufactured at reasonable cost, it is widely used as a fertilizer on both agricultural and forested lands. It is normally applied as small pellets or prills. The urea will react with water in the soil to release the nitrogen for plant uptake. This photo shows a helicopter dropping white urea pellets to fertilize a forest.

Volcanic Ash

Small glassy particles high in silica that are ejected during a volcanic eruption. The molten rock ejected by the volcano rapidly cools in the air so crystals don't form and the solidifying material is glassy. Air bubbles are trapped in the glassy particles as the ash moves through the air before falling to the land surface. This scanning electron micrograph shows the vesicular shapes often found in the ash where air bubbles have broken. These ash particles have the size of fine silt (less than 0.05 mm but much bigger than 0.002 mm). This ash is from Mt. St. Helens.

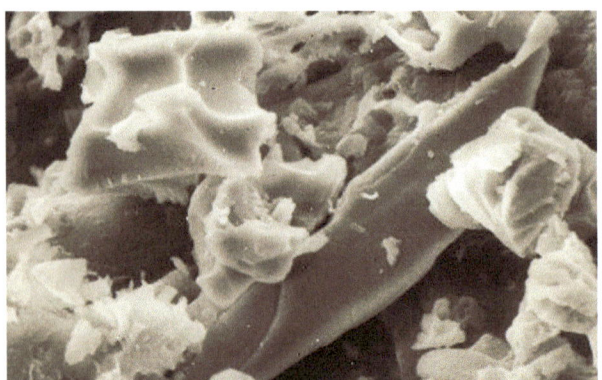

Watershed

An area of land where all precipitation flows into a basin and subsequently into a waterway such as a stream or river. This picture shows an outlined watershed that extends from a glacier to a marine inlet in British Columbia. Note that much of the upper watershed is hidden behind mountains.

Weathering

Most minerals and rocks are chemically unstable at the temperatures and pressures at the surface of the earth. This means rocks can be broken down into their *mineral* components, and minerals can be broken down to individual atoms or ions of elements. Weathering can be physical (breaking a bigger rock into smaller pieces) or chemical (changing the chemical composition by removing or adding elements or ions). The example here shows physical and chemical weathering of granite. The photo above is of an unweathered piece of granite; note that the mix of small individual mineral crystals gives the rock a speckled appearance (these minerals are mostly quartz, feldspar, and mica; photos of large crystals of these three minerals can be seen with the definition of *mineral*). The photo below shows granite after it has undergone chemical weathering which changes its color (iron from the mica has formed oxides making the rock more orangish) and physical weathering which is breaking the rock into individual crystal minerals. Weathering causes minerals from the original rock to disappear but provides the building blocks for new oxides, clays, and other minerals to form in soil. Scale is in cm.

Appendix 2: List of Sources/Further Reading

Chapter 1

Hillel D (1991) Out of the Earth, Civilization and the life of the soil. University of California Press, 321 p.

Jenny H (1941) Factors of Soil Formation. A system of Quantitative Pedology. 1941. Republished in 1994 by Dover Publications, New York. 281p.

Swaisgood RR, Sheppard JK (2010) The culture of conservation biologists: show me the hope. BioScience 60(8):626–630.

Chapter 2

Keam RF (1988) Tarawera. The Volcanic eruption of 10 June 1886. A comprehensive account. 472p. ISBN 0-473-00444-5.

Chapter 3

Kiple KF, Orne KC (eds) (2000). The Cambridge World History of Food. II.A.7. – Rice. Cambridge University Press. www.cambridge.org/us/books/kiple/rice.htm.

Liu F, Chen S, Peng J, Chen G (2012) Temporal variations of water discharge and sediement load in Huanghe River, China. Chin. Geogra. Sci. 22(5):507–521. doi: 10.1007/s11769-012-0560-y. www.springerlink.com/content/1002-0063

Rengasamy P (2006) World salinization with emphasis on Australia. Journal of Experimental Botany Plants and Salinity Special Issue. 57(5):1017–1023. doi:10.1093/jxb/erj108.

Chapter 4

Binkley D, Fisher R (2013) Ecology and Management of Forest Soils, 4th Edition Wiley-Blackwell, 362 p. ISBN: 978-0-470-97946-4.

Kimmins JP (2004) Chapter 11, Soil: The Least Renewable Physical Component of the Ecosystem. In: Forest Ecology: A foundation for sustainable forest management and environmental ethics in forestry. Prentice Hall/Pearson Education, New Jersey. 611p.

Chapter 5

Buol S (2008) Soils, Land, and Life. Prentice Hall, 320 p.

Magdoff F, van Es H (2010) Building Soils for Better Crops, 3rd Edition. Sustainable Agriculture Research and Education (SARE). 294p.

Chapter 6

Dickson JL, Head JW, Levy JS, Marchant DR (2013) Don Juan Pond, Antarctica: Near-surface $CaCl_2$-brine feeding Earth's most saline lake and implications for Mars. Nature Scientific Reports 3: 1166, DOI: 10.1038/srep01166. http://www.nature.com/srep/2013/130130/srep01166/pdf/srep01166.pdf

Chapter 7

Ashton BC (1933). The Napier Ahuriri Lagoon Lands. In R.H. Hooper (Ed) The New Zealand Journal of Agriculture pp 69–77 and 260–266. New Zealand Department of Agriculture.

Daly BK, Rijkse WC (1976) Saline soils of the old Ahuriri Lagoon, Napier, New Zealand. Soil Bureau Scientific Report 27.

Department of Survey and Land Information (1989) 100 years ago [cartographic material] : plan showing Ahuriri Lagoon, Scinde Id. & surroundings up to and at 1865 : 100 years of progress : the same area after reclamations and earthquake action 1965. Dept. of Survey and Land Information, Wellington, N.Z.

McSaveney E, Nathan S (2013) 'Geology – overview - Holocene – the last 10,000 years', Te Ara – the Encyclopedia of New Zealand, updated 4-Sep-13 URL: http://www.TeAra.govt.nz/en/map/8406/uplift-of-new-zealand

McSaveney E, (2012) 'Historic earthquakes – Rebuilding Napier', Te Ara – the Encyclopedia of New Zealand, updated 13-Jul-12 URL: http://www.TeAra.govt.nz/en/historic-earthquakes/page-8

Gooley T (2012) The Natural Explorer. Understanding your landscape. Hodder and Stoughton Ltd, London. 360p.

Rosser BJ, Ross CW (2011) Recovery of pasture production and soil properties on soil slip scars in erodible siltstone hill country, Wairarapa, New Zealand. New Zealand Journal of Agricultural Research 54(1):23–44.

Tonkin and Taylor Ltd (2013) Liquefaction Vulnerability Study. Report Prepared for the Earthquake Commission. Sourced at http://www.eqc.govt.nz/sites/public_files/documents/liquefaction-vulnerability-study-final.pdf

Xiwei Xu, Wenbin Chen, Wentao Ma, Guihua Yu, and Guihua Chen 2002. Surface Rupture of the Kunlunshan Earthquake (M_s 8.1), Northern Tibetan Plateau, China Seismological Research Letters, November/December 2002. 73:884–892. DOI 10.1785/73.6.884.

Chapter 8

Blume, HP, Leinweber P (2004) Plaggen Soils: landscape history, properties, and classification. Plant Nutr. Soil Sci. 167:319–327. DOI: 10.1002/jpln.200420905

English Heritage (2011) Introductions to heritage assets. Prehistoric Henges and Circles. Accessed at www.english-heritage.org.uk/publications/iha-prehistoric-henges-circles/prehistorichengesand-circles.pdf www.stonecircles.org.uk

Phillips WJ, revised by Huria J (2008) Maori Life and Customs. Penguin Group (NZ) Ltd. 191 p. ISBN 9780143009726.

Romero, F. (2011) Asase Yaa. From Top 10 Earth Goddesses, Time. Accessed at http://content.time.com/time/specials/packages/article/0,28804,2066721_2066724_2066705,00.html

University of Georgia 2000. Creation Stories from around the World. Encapsulations of some traditional stories explaining the origin of the Earth, its life, and its peoples. Fourth Edition July 2000 Sourced at http://www.gly.uga.edu/railsback/CS/CSIndex.html

Chapter 9

Cone, M (2009) Should DDT be used to combat malaria? Scientific American. May 4 2009. http://www.scientificamerican.com/article/ddt-use-to-combat-malaria/

Global population statistics: http://www.census.gov/ipc/www/idb/worldpopinfo.html

Hillel, DJ (1990) Out of the Earth, civilization and the life of the soil. The Free Press, New York. 321p.

New Zealand Historic Places Trust. Rotowaro Carbonisation Plant. http://www.heritage.org.nz/the-register/details/7013

Tarchitzky, J (Undated) United Nations Department of Economic and Social Affairs Report CSD16/17 on Agriculture in Israel. Sourced at www.un.org/esa/agenda21/natlinfo/countr/israel/agriculture.pdf

Stirzaker R (2010) Out of the Scientists Garden. A story of water and food. CSIRO Publishing, Canberra. ISBN 9780643096585.

Appendix 3: Photographic Credits

Sources of photographs that are not those of the authors are acknowledged in the relevant figure captions and reproduced with permission of the photographers who reserve all rights to their use.

Photographs and diagrams for which no acknowledgment is included in the text are those of either Megan and Errol Balks or Darlene Zabowski. They are provided for reproduction in this publication, all rights reserved.

Figures 9.31 and 9.32 and cover illustrations are reproduced with permission of Marianne Coleman and Megan Balks, all rights reserved.

Appendix 4: About the Authors

The authors are two women from opposite ends of the Earth who are passionate about soils and soil science. They like playing with mud and getting their hands dirty!

Dr. **Megan Balks** has over 30 years of experience in soil-related study, research, and teaching. Megan is based in New Zealand (at the University of Waikato), and her experience includes 19 trips to study soils in Antarctica as well as work throughout New Zealand. Megan has also traveled widely to study soils having undertaken fieldtrips in many parts of the world, including the Arctic (Russia, Norway, and Alaska), Australia, Thailand, Peru, China, Samoa, and Europe. Megan undertook her PhD studying irrigation of effluent onto land, and she has also worked on water irrigation schemes in Otago and effluent irrigation in Australia. She has supervised over 40 graduate thesis projects on the widest range of soil-related studies from soil fertility in New Zealand hill country, through irrigation of city wastewater onto land and study of some of the most southerly soils on the planet in Antarctica. Megan is a fellow of the New Zealand Society of Soil Science (the first woman to receive that honor). With her husband Errol, Megan owns a small hill country sheep farm, which also includes about 60 acres of New Zealand native forest, and so she has hands-on experience in managing the land.

Professor **Darlene Zabowski** has a BS in forest ecology and an MS and PhD in soil science. She worked as a research soil scientist for the US Forest Service before transferring to the University of Washington in 1993. She has conducted research on soils and taught introductory soils and advanced classes in soil science for 30 years. She has received several awards for excellence in teaching. Darlene has worked with many soils in various areas of the USA, as well as Canada, New Zealand, and China, and participated in field trips in many other parts of the world. Her research has mostly focused on forest soils but has often included comparative research with soils from other ecosystems and a variety of landscapes. Darlene is an avid hiker and enjoys keeping the soil in her vegetable garden healthy.